NEUROLOGY - LABORATORY AND CLINICAL RESEARCH DEVELOPMENTS

COGNITIVE IMPAIRMENT AND SURVIVAL AFTER A NATURAL DISASTER

LESSONS LEARNED FROM LIFE EXPERIENCES IN THE GREAT EAST JAPAN EARTHQUAKE 2011

NEUROLOGY - LABORATORY AND CLINICAL RESEARCH DEVELOPMENTS

Additional books in this series can be found on Nova's website under the Series tab.

Additional e-books in this series can be found on Nova's website under the eBooks tab.

NATURAL DISASTER RESEARCH, PREDICTION AND MITIGATION

Additional books in this series can be found on Nova's website under the Series tab.

Additional e-books in this series can be found on Nova's website under the eBooks tab.

NEUROLOGY - LABORATORY AND CLINICAL RESEARCH DEVELOPMENTS

COGNITIVE IMPAIRMENT AND SURVIVAL AFTER A NATURAL DISASTER

LESSONS LEARNED FROM LIFE EXPERIENCES IN THE GREAT EAST JAPAN EARTHQUAKE 2011

KENICHI MEGURO

nova biomedical
New York

Library of Congress Cataloging-in-Publication Data

Names: Meguro, Kenichi, author.
Title: Cognitive impairment and survival after a natural disaster : lessons learned from life experiences in the Great East Japan Earthquake of 2011 / Kenichi Meguro (Geriatric Behavioral Neurology, Tohoku University, Seiryo-machi, Sendai, Japan).
Description: Hauppauge, New York : Nova Science Publisher's, Inc., [2017] | Series: Neurology-laboratory and clinical research developments | Series: Natural disaster research, prediction and mitigation | Includes bibliographical references and index.
Identifiers: LCCN 2016059863 (print) | LCCN 2017006592 (ebook) | ISBN 9781536108248 (softcover) | ISBN 9781536108361 (ebook) | ISBN 9781536108361
Subjects: LCSH: Cognition disorders. | Disasters--Psychological aspects. | Survival. | Tohoku Earthquake and Tsunami, Japan, 2011.
Classification: LCC RC553.C64 M44 2017 (print) | LCC RC553.C64 (ebook) | DDC 616.8--dc23
LC record available at https://lccn.loc.gov/2016059863

Published by Nova Science Publishers, Inc. † New York

Contents

INTRODUCTION

Behavioral neurology is a subspecialty of neurology that investigates the neurological basis of human behavior and cognitive function. The associations between neurological diseases, cognitive function and treatment are also the foci of investigation. Recently, dementing diseases, such as Alzheimer's disease, have become so common that their behavioral abnormalities are now being investigated with reference to brain damage.

During a big disaster, which can critically affect our lives, we should utilize our "brains" fully in order to survive, i.e., to understand the situation around us, to make a proper judgement, and to decide our behaviors; all of which are associated with brain functions. Figure 1 illustrates the concept for brain-behavioral relationship during the disaster, and Figure 2 shows the brain functions associated with behavior during the disaster 1: understanding the situation. Understanding the situation primarily requires the posterior part of the brain, especially the parietal lobe.

Briefly, the occipital (shown by green, Figure 2), temporal (pink), parietal lobe (blue), and frontal lobes are related to visual, auditory, other sensory, and motor functions, respectively. Each lobe has primary and secondary areas; the former is associated with primary function, whereas the latter is related to the association functions, which add "meaning" to the primary information. In particular, visual and auditory information should be fully integrated to understand the situation and make judgment, which is the function of the parietal lobe.

However, in my opinion, "disaster medicine" is focused mainly on the environment, such as infectious diseases (which are of course important), with the underlying assumption that our brain functions are always "normal" and thus judgement and behavior remain "normal." By contrast, "dementia medicine" does not always consider dementia in determining peoples' behaviors during the disaster. I believe that approaches to "brain-based disaster medicine" and "disaster-associated dementia medicine" have not been fully integrated nor even investigated.

Herein, I focus on reconstruction support after the Great East Japan Earthquake 2011, from an academic perspective, with particular respect to behavioral neurology based on my own experiences.

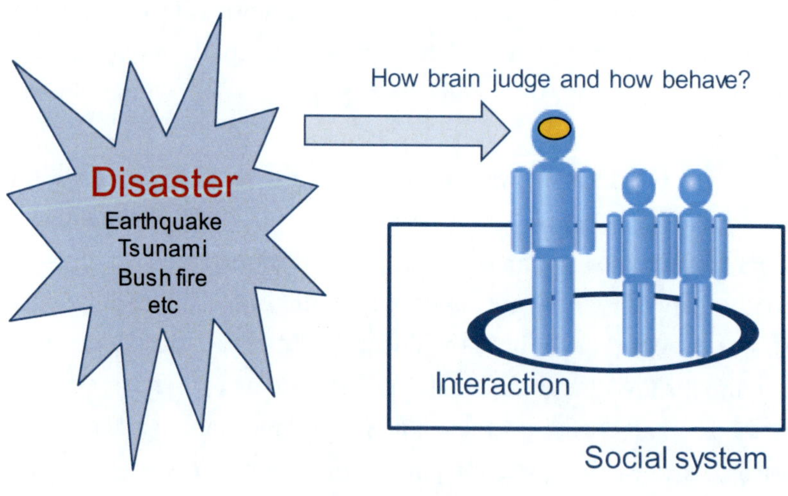

Figure 1. Brain-behavioral relationship under the disaster.

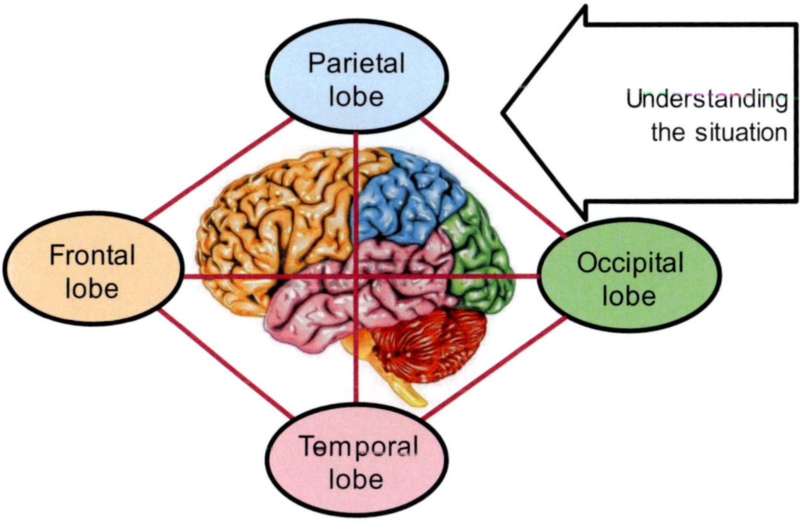

Figure 2. Brain functions associated with behavior under the disaster 1: understanding the situation.

Chapter 1

WHAT WAS THE GREAT EAST JAPAN EARTHQUAKE 2011?

The goals of my research group, i.e., Geriatric Behavioral Neurology, Tohoku University, are to study the neuroscience of aging and dementia and to contribute to the welfare of elderly people in the community. Our priority has always been our work in the northern area of Miyagi Prefecture, i.e., the cities of Osaki-Tajiri (Meguro et al., 2002; 2003; 2004; 2007; 2014a; 2014b; Ishii et al., 2007; 2009; Yamaguchi et al., 2009), Kurihara (Meguro et al., 2012a; 2012b; Kasai et al., 2012; Nakamura et al., 2013; Takada et al., 2014; Kumai et al., 2016; Kobayashi et al., 2016), Tome (Akanuma et al., 2016), and the town of Wakuya. However, I never expected to experience such an extreme challenge as that of the Great East Japan Earthquake 2011.

The Great East Japan Earthquake 2011 occurred at 14:46 on March 11, 2011. The magnitude was 9.0, with a maximum intensity of 7.0 on the Japanese scale. The epicenter was 130 km off the coast, as illustrated in Figure 3 (Satomi, 2011). The mega tsunami that followed the earthquake caused devastating damage to the coastal areas. According to the Reconstruction Agency (February, 2015), more than 160,000 people are still living in temporary housing, as shown in Table 1. Despite the availability of good housing for younger people, isolation remains a problem for older adults

Table 1. The present situation and temporal housing

	numbers
Deceased (+ Missing)[1)]	18,456
Refugee[2)]	app. 97,000
Temporary housing res.[3)]	33,748 households

[1)] February 2016, by National Police Agency
[2)] May 2017, by Reconstruction Agency of Japan, May 2017.
[3)] January 2017, by Kahoku Newspaper

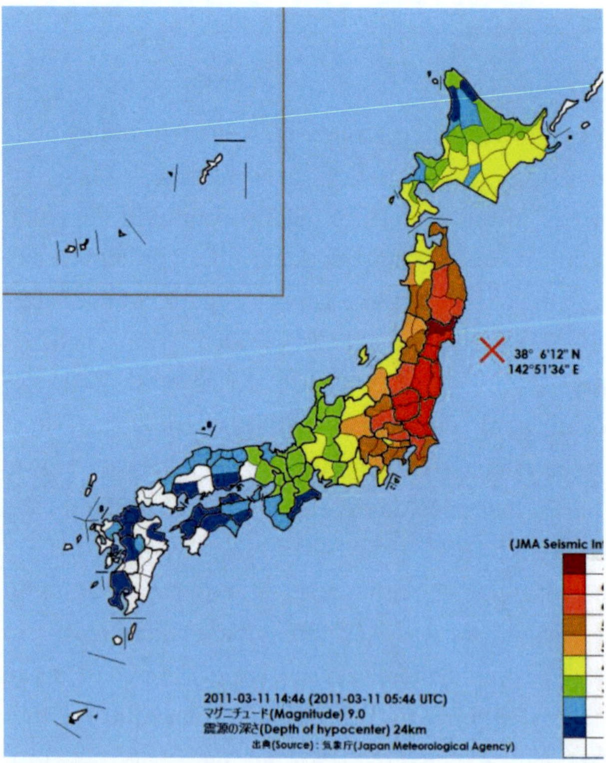

Figure 3. Great East Japan Earthquake 2011. *Satomi S. Surg Today 2011;41:1171-1181.*

HISTORICAL PERSPECTIVE

On the first day of the disaster, I was seeing patients in my research field (see Chapter 2). I was not able to understand what had happened since we were in a blackout with no TV available during the electric outage. Several days later, while I saw the TV news on the disaster, I still did not fully understand the situation. After obtaining the historical information on the earthquake, I eventually understood what was happening. As described below, it was a repeated earthquake over a 1,000-year span. From my experience I should say that this kind of "historical orientation" is critical in understanding the situation.

As Figure 4 shows, the earthquake had been predicted by science, i.e., seismic stratigraphy (Normile, 2011). The sediment from the previous tsunami in the 9th century (869), which was supported by a description in an old document, predicted that a similar earthquake would occur about 1,000 years later. However, despite the scientific warnings, nobody had considered that this kind of 1,000-year level event could occur tomorrow. As the adage "history repeats itself" suggests, an old document from the 9th century (Fujiwara et al., 869) recorded the same earthquake and tsunami, as shown in Table 2.

As another proof that "history repeats itself," an old shrine from the 9th century, the "*Nami-wake*" shrine (Figure 5), which means the "separation (*wake*) of wave (*nami*)," has been located, and the modern day tsunami arrived at the exactly the same point! Lord of Sendai-*han* (Sendai Domain), *Masamune Date* (1567-1636) studied the historical records, and thus located the downtown at the middle of the river of *hirose*, which is an unlikely location compared with other historical castle towns (e.g., Osaka, Hiroshima, Edo (Tokyo)), which were located at the downstream end of rivers, as apparently effective deterrent to protect against enemies.

Figure 4. Great East Japan Earthquake 2011 predicted by science: seismic stratigraphy. *Modified from Active Fault and Earthquake Research Center, News No. 16, August 2010; Normile D, Science 2011;332:22-23.*

Table 2. History repeats itself: an old document in 9C recorded the same earthquake and tsunami

陸奥国地大震動。流光如晝隠映。頃之。 人民叫呼。伏不能起。或屋仆壓死。 或地裂埋殪。馬牛駭奔。或相昇踏。 城郭倉庫。門櫓墻壁。頽落顛覆。不知其数。 海嘯哮吼。聲似雷霆。驚濤涌潮。泝徊漲長。 忽至城下。**去海数千百里。** 浩々不辨其涯涘矣。原野道路。惣為滄溟。 乗船不遑。登山難及。**溺死者千許。** 資産苗稼。殆無子遺焉。 三代実録（日本紀略、類聚国史） 貞観11年5月26日（July 13, 869）

A big earthquake occurred at the *Mutsu* country (Tohoku district). People cried, were not able to stand up. A big tsunami came and thousands of people were killed.

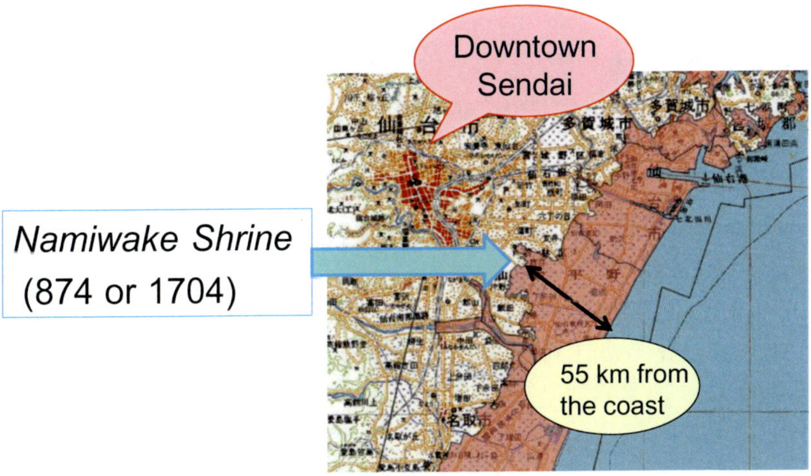

Figure 5. History repeats itself: an old shrine in 9C located in the same arrival point of tsunami. "*Nami-wake*" shrine, which means the "separation (*wake*) of wave (*nami*)." *Modified from Geospatial Information Authority of Japan, 2011.*

Lessons Learned

- We should be modest when learning from history, since "history repeats itself." An old document from the 9th century recorded the same earthquake and tsunami, and an old shrine was built as a marker of the previous arrival point of tsunami, which was almost the same as the current tsunami landfall.
- I should indicate that this kind of "historical orientation" is critical to understand the situation.

Actual Status after the Earthquake

At that time a figure published in Lancet (McCurry, 2011) indicating that downtown Sendai was involved in the tsunami caused misunderstanding among the readers. I was asked by my friends abroad whether Tohoku University was floating on the sea. Figure 6 shows the figure and the photos demonstrating that Tohoku University and my office at that time were intact, in contrast to the figure published by the journal.

The TV news at that time repeatedly showed "shocking" tsunami scenes and thus people outside of the Tohoku district hesitated to enter the damaged areas, even into downtown Sendai. Figure 7 shows photos taken by myself to inform people outside my city that the citizens' daily lives had started to recover 9 days after the earthquake.

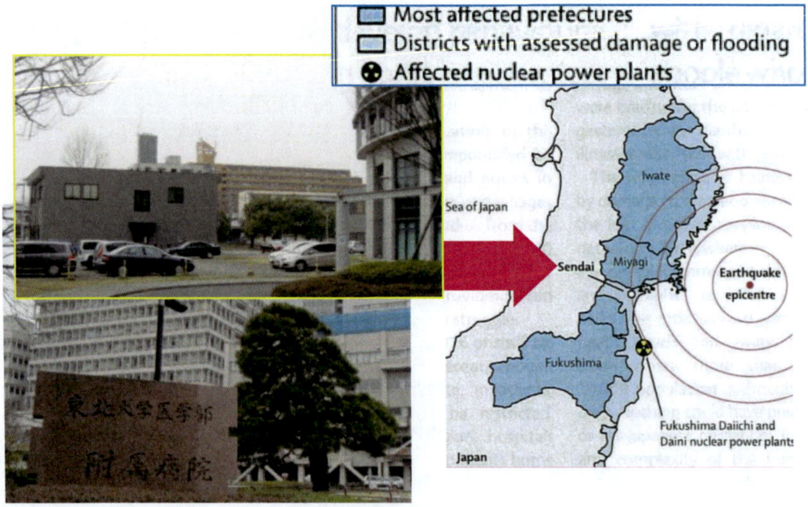

Figure 6. Lancet figure's error causing misunderstanding. *Modified from McCurry J. Lancet 2011;377:1061-1062. Photos by author.*

Figure 7. Downtown Sendai, March 20, 2011 (9 days after). *Photos by author.*

NEGATIVE EFFECT OF MASS MEDIA: POSSIBLE PTSD AND "NEGATIVE" PERSONAL REMINISCENCE APPROACH

Exposure to natural disasters through the media is widespread. This phenomenon of indirect exposure actually reveals that the disaster is a type of "brain" problem. Although exposure to trauma has a wide range of psychopathological consequences, posttraumatic stress disorder (PTSD) has been shown to be the most common. Evidence from epidemiologic studies indicate initial but not extended increases in the rates of PTSD symptoms among individuals exposed indirectly through mass media (Neria et al., 2011); consequently, the Japan Medical Association requested TV stations to refrain from broadcasting such shocking images.

From my perspective, TV images and information from the primary stage, as well as my actual visits to the damaged areas during the secondary reconstruction stage, were extremely stressful and "negative" reminiscence approach on me. The completely "dead" scene of the coastal area where I learned to swim during my childhood, that of the islands where I visited for biology field trips during my high school years, etc., negatively affected my remote memory in the brain.

SPECIAL THANKS TO TAIWAN

During that difficult time, so many countries sent rescue to aid the Japanese people, as illustrated in Figure 8. The total contribution was 34 billion yen. As a Japanese person, I really appreciate their kindness. In particular, the people of Taiwan contributed more than half of the total sum, corresponding to 1,500 Yen per person (from babies to the elderly).

Figure 8. International rescue team. *Modified from Ministry of Foreign Affairs, Japan, 2011.*

UNIQUE CHARACTERISTICS OF THIS EARTHQUAKE

Few houses collapsed during this earthquake because of the lessons learned in the Miyagi Oki Earthquake of 1978, when many people died due to the collapse of houses. Based on a prediction of similar earthquakes to likely occur every 30 to 40 years, we had taken preventive measures. This also resulted in fewer seriously injured patients, as fewer people were crushed, etc. However, the damage from the tsunami was still devastating.

Figure 9. Shortage of medical supplies, food, daily commodities. Everything! *Photos by author.*

The Disaster Medical Assistance Team (DMAT) came to support from around Japan; however, few cases rescued from debris required acute medical treatment. Nearly all of those caught by the tsunami were killed, while those who could escape it were at least able to walk: they were either healthy or dead. Instead, treatment as part of the chronic stage was required (Satomi, 2011).

The disaster caused panic at several backup hospitals due to the sudden increase of patients. These patients were suffering from the lack of treatment for chronic illness due to the destruction of hospitals near the coastal areas. There were also problems as a consequence of poor hygiene conditions in most emergency shelters.

Transportation was paralyzed due to shortage of gasoline by the emergency shutdown of oil factories. Thus, there was a shortage of medical supplies, food, and daily commodities; everything! Figure 9 shows the long line of cars waiting for gasoline.

Effects on Chronic Diseases Reported

The effects of the sudden discontinuation of treatment for chronic diseases were reported. For hypertension, home blood pressure was significantly elevated immediately after the tsunami and decreased gradually over the following month (Satoh et al., 2011). For diabetes mellitus, disease control was worse in the tsunami-affected patients compared with those affected by the earthquake alone; especially among women (Ogawa et al., BMJ Open 2012). With respect to arrhythmias, the number of patients experiencing tachyarrhythmias and hospitalization of patients with implanted cardiac devices exhibiting heart failure increased (Nakano et al., 2012). Regarding home oxygen therapy, 20% experienced worsening due to tracheobronchial infection and interruption of treatment. A temporary shelter for patients with oxygen therapy was established (Japanese Red Cross Ishinomaki Hospital) (Kobayashi et al., 2012).

For Alzheimer's disease, alterations of the Mini-Mental State Examination (MMSE) (Folstein et al., 1975) and Neuropsychiatric Inventory (NPI)-Q were reported after the earthquake and tsunami. A comparative study (Furukawa et al., 2012) among Group 1 (who did not suffer from the disaster), Group 2 (subjects who had experienced the disaster but did not stay at an evacuation site), and Group 3 (subjects who were forced to stay at an evacuation site), disclosed that the MMSE scores of Group 3 decreased compared with those of Groups 1 and 2. The NPI-Q scores increased step by step through Groups 1, 2, and 3.

Lessons Learned

- The actual status after the earthquake was different from that published in famous medical journals or in the shocking TV images broadcasted repeatedly, which might have negative effect. However, actually visiting the damaged area may have an effect of "negative" personal reminiscence approach.
- The essence of this disaster was sudden paralysis of the transportation system due to a shortage of gasoline following the emergency shutdown of oil factories. Thus there was a shortage of medical supplies, food, and daily commodities; everything!
- Medically, the effects of the sudden discontinuation of treatment for chronic diseases were reported.

Chapter 2

MY LABORATORY'S EFFORTS DURING THE PRIMARY EMERGENCY STAGE AND THE SECONDARY SUPPORT STAGE

I will repeat that the goals of my research group, i.e., Geriatric Behavioral Neurology, Tohoku University, are to study the neuroscience of aging and dementia and to contribute to the welfare of the elderly people in the community. Our work in the northern area of Miyagi Prefecture has always been our priority. I encouraged my staff and myself that this was the time to do our best to support the areas.

PRIMARY EMERGENCY STAGE

I saw patients in Kurihara, where the maximum intensity of 7.0 on the Japanese scale was recorded. Our initial goals were to protect patients from moving objects and evacuating the building (Meguro, Neurology 2011). We immediately ensured the safety of the patients by evacuating the building. When nurses gave blankets for them outside the building, I saw a dementia patient call out to the nurses, "Not for me, but for the patients first!" He had been a fireman before retirement. His remote memory was apparently stimulated and he was attempting to help other people.

All the staff knew how to act because of their experience of the 2008 Iwate-Miyagi Inland Earthquake, which hit this city only 3 years ago. Our initial goals were to protect ourselves and our patients from moving objects and then evacuating everyone from the building.

After ensuring that everything was in order at the clinic in Kurihara City, I went to the Osaki-Tajiri SKIP Center, which is one of our most important research centers (Meguro et al., 2002). This is an integrated institute for medicine and welfare for the prevention of stroke, dementia, and bed confinement among the elderly. I found that a large number of refugees had gathered at the Center. However, as the doctor in charge was off duty that day, an otolaryngologist was the only doctor available, and he had sustained a bone fracture in the earthquake. While one of our graduate students transported the injured doctor to an emergency hospital, I took on the duties of the clinic, confronting a series of challenges caused by the disaster.

Firstly, I started by confirming whether they carried their medications. When I was the president of the center during previous disasters, I always informed the staff who were going to help victims at other locations as follows: "The victims would come to shelters with valuables, such as money, next to the skin. However, the medications were also considered valuables, especially for those elderly with chronic diseases. Sudden cessation of some drugs, such as anti-parkinsonian drugs or antipsychotics, could cause adverse effects; thus the staff should first confirm whether they had brought their medication." Thanks to the staff training, I was pleased to find that most of the 80 refugees had carried their own medications with them, which enabled us to make up a list of their doctors and medications. While some patients did not have their medications and had experienced angina or worsening of diabetes mellitus, all of these were managed successfully. None of the patients went without their antipsychotic or antiparkinsonian drugs; the absence of which might have been troublesome (Rabinak et al., 2010; Baldessarini et al., 2010). Among older people, suspected dementia patients were confused by unfamiliar shelters, leading to some troubles with healthy people.

An important lesson in disaster preparedness is that it is essential to repeatedly teach everybody, including the members of the Social Welfare Council, about the critical importance of medications.

After restoring our satellite office at Tajiri, I confirmed that my family was safe and traveled to the university laboratory in Sendai City. There was initial misunderstanding about the location of the University-the central part of Sendai City where the university is located is far from the coast, and it was completely safe. All the members of our laboratory and their families were fortunately confirmed as being safe within 1 week. After restoring our laboratory office, we were soon able to resume our usual duties.

At a meeting of the headquarters for emergency disaster control of our Graduate School of Medicine, the Dean Professor Yamamoto was encouraging: "Conduct your daily business-that is restoration!" Our team members worked day and night at our collaborative facilities. After getting permission for the use of 3 cars as emergency vehicles, which were made available by 2 staff members and a graduate student, we discussed possible support for refugee shelters that might ask for our help. In preparation for a disaster, we should know how to make our cars act as emergency vehicles.

Secondary Support Stage

Ten days after the earthquake, the "primary emergency stage," in which the activities of the rescue and emergency teams play the most critical role, was followed by the "secondary support stage." During this time, refugees from the most damaged areas began to arrive at the less damaged areas, including our own research areas of Kurihara City, Tome City, and Osaki City.

Many people, especially those who had graduated from my university, approached us to help as volunteers. I established my policy as follows:

1. Ensure the safety of the destination area. Some refugee shelters were situated near the coast, with a risk of secondary incidents.
2. Identify the probable demands on the destination area so that preparations could be made before people arrive. As it happened, 40 volunteer medical doctors were waiting at a shelter with nothing to do, having rushed to the site to help refugees, without any plan in place.
3. Understand the goals of volunteers and try to match their goals with needs, taking into account any potential dangers, including psychological trauma. Some volunteers or students could be so eager to help that they might not be mindful of danger.

Following establishment of these policies, I contacted our research centers and received 2 requests: Osaki City requested help in confirming the safety of people in the community, and helping with the mental care of people with psychological trauma. Both locations needed help supporting the shelters.

At that time, three vehicles belonging to my staff were registered as "emergency cars" and attempted to help victims in the areas. However, we faced the Ministries and agencies' powers in their areas of jurisdiction. My staff went to the National Police Agency in order to register. However, the information regarding the lack of gasoline and which station we should go to in order to supply the cars were within the jurisdiction of the Ministry of Transport. They then had to go another office. The information on the highway was within the purvey of the Ministry of Land, Infrastructure, Transport and Tourism, which meant they also had to visit another office. The information on the WEB did not support the real situation for all, and easily became confusing; however, we did our best.

As described in Chapter 1, Lord of Sendai-han (Sendai Domain), *Masamune Date* (1567-1636) studied the old history, and placed the downtown at the middle of the river. He was a "dictator" standing above the Ministries and thus successfully made the city safe at that time. I will not attribute the problems we experienced to democracy; however, a simple and urgent system to unify information and decision making during such a risky condition should be considered.

Activities of the Support Team for Osaki City

Seven members, including 3 graduate students, 2 former members, and 2 staff members, joined a team that worked from March 20-25. Following our request to the Committee for Disaster Restoration of the university for more help, 4 additional team members were dispatched (Figure 10).

Our team visited 790 houses and discovered 54 people who were in need of help. Our visits accounted for 10.4% of the 7,604 houses that were visited in Osaki City. We discovered various cases of "weakness due to the disaster." Most of the people who needed help were elderly, including some who had become sick. The shortage of fuel had prevented some people from consulting their doctor. Others were confused because they did not have the information that they needed to obtain medicines from a doctor other than their usual consulting physician. Some elderly people who lived alone were found deceased in their house. We discovered various cases of "weakness due to the disaster." Most of the weakness due to the disaster was previously assessed as Clinical Dementia Rating (CDR) 0.5 in the Tajiri Project (Hughes et al., 1982; Morris et al., 1993; Meguro et al., 2004), which is the borderline state between healthy aged and dementia. It is important to routinely screen CDR 0.5 residents in a community. Our contributions may help in the planning of future health policies.

A psychiatrist, one of the former members of our research team, visited shelters in Matsuyama Town and in Osaki City. In contrast to the primary emergency stage, the situation seemed to be moving to the next stage of restoration, in which many refugees were beginning to go back to their homes. Some people stayed at the refugee centers, including many elderly people who had been living alone, or had psychiatric disorders, dementia, or other kinds of aging-related illness. Although some refugees were not directly damaged by the earthquake, the system of Long-Term Care Insurance was important, especially for elderly persons (Takeda et al., 2010). The efforts of the Social Welfare Council care managers enabled elderly refugees to be sheltered efficiently in the facilities, to receive home help services, and to receive delivery of medications. Some psychiatric patients might not have received proper treatment, with worsening symptoms in the face of such a disaster.

We should consider that acute traumatic stress may be a normative response to a life threat, which tends to subside once conditions of safety are established. However, during the "secondary support stage," there is a residual minority of survivors who will come to experience chronic posttraumatic stress disorder and their deeds can be easily overlooked (Silove et al., 2006).

Figure 10. Activities of the support team for Osaki City. Requested help confirming the safety of people in the community. *Photo by author. Meguro. Neurology 2011;77:e12.*

Activities of the Support Team for Tome City

Since March 27, the shelters in Tome City have accommodated 2,500 refugees. We dispatched a team from our laboratory comprised of a psychiatrist and a psychotherapist. The activity base was Tome City Sanuma Hospital, at which one of our research fellows was based. He had been working with us in a collaboration among hospitals and administrations, which this served us well for efficient action after the earthquake. Some refugees with psychotic symptoms were admitted to the hospital. The psychiatrist, one of our research fellows, took on regular consultations for the psychiatric patients, and this work is likely to continue as part of the long-term support for refugees.

Figure 11. Activities of the support team for Tome City. Medical support of shelters with permission from the principal of Tome Junior High School. *Photo by author. Meguro. Neurology 2011;77:e12.*

At the shelters, each refugee needed advice about health-related concerns (Figure 11). In order to respond to this demand, a psychotherapist, a former member of our research team, served as a volunteer to provide health advice and help with sorting of the medications. Most of the refugees in Tome City had lost their houses. While they benefited from spending time with their neighbors, many experienced great stress, possibly related to the lack of privacy. In the Tome City shelters, the situation appeared to remain at the primary emergency stage, rather than moving on to the next stage of restoration.

Table 3 summarize the activities of my research team during the disaster.

Table 3. Activities of my research team under the Great East Japan Earthquake 2011

	contents	total number of staff	targets	total number of targets	period	place
1	rescue and waqtch at shelters	2	refugee	160	March 11-12	Osaki-Tajiri SKIP Center
2	transport victims to emergency hospital	2	injuried patients	2	March 11-12	Osaki-Tajiri SKIP Center
3	confirming the safety of people in the community and support shelters	22	residents and refugee	820	March 20-25	Osaki City
4	transport relief supplies	12	instituionalized patients	120	March 26-April 1	Cities of Osaki and Tome, Neurisng homes an dhospitals
5	mental care of victims and support chelters	12	refugee	1,800	March 22-about 1 year	Tome City
6	medical care for inpatients from the damaged areas	8	inpatients	80	April 3-24	Kawasaki Kokoro hopsital
7	medical care for instituionalzed patients from the damaged areas	3	institutionalized patients	24	April 24-25	Kesen-numa City
8	transport relief supplies	3	residents	30	April 24-25	Kesen-numa City
9	transport relief supplies	6	institutionalized patients	320	April 28-May 12	Nursing homes and hospital

Lessons Learned

- Some drugs, such as antipsychotic or antiparkinsonian drugs, can have adverse effects if they are suddenly stopped. An important lesson in disaster preparedness is to repeatedly teach everybody, including the members of the Social Welfare Council, about the critical importance of the medications.

- Confirming the safety of residents revealed that most of the weakness due to the disaster was previously assessed as CDR 0.5. Cognitively healthy people, i.e., Clinical Dementia Rating (CDR) 0 were able to escape to shelters by themselves, and patients with dementia (CDR 1+) were aided by the officials or public health nurses and were taken to specific shelters. However, residents with borderline states between healthy and dementia, i.e., mild cognitive impairment (MCI), CDR 0.5, experienced problems, as was previously found in the Osaki-Tajiri Project. These individuals could not understand broadcast information regarding the disaster and became confused due to the limited services. In particular, during the hyperacute stage of the earthquake, these people should have stayed in a safe place or escaped outside in anticipation of the aftershocks. However, it was reported that some did not think about the aftershocks and manifested inappropriate behaviors (e.g., checking rooms in their home, repositioning objects, taking care of graves near the house, and calling the police).

Chapter 3

CHRONIC EFFECT ON DEMENTIA

EFFECT ON DEMENTIA IN A NURSING HOME

At that time, March 11, it was still cold in northern Japan. Heating and quilts were not provided to many nursing homes, and patients with dementia were cared for with the minimum requirements, using heavy blankets to keep themselves warm. This meant no physical rehabilitation was possible for them and thus their levels of activities of daily living (ADL) decreased. Furthermore, the shortage of fuel prevented the staff from consulting their doctors.

No electricity was supplied to nursing homes. The staff had to care for dementia patients according to the natural rhythm. Namely, they awakened the patients at sunrise, and put them to sleep at sunset. As a result, in some nursing homes, the sleep/wake rhythm was actually improved in some patients.

Usually, the patients were cared for individually by staff in what is called "unit care." However, the specific circumstances during the disaster, in which the staff could not focus on such intensive care, forced them to conduct "mass care"; this might exert some effect of "group care."

One Case Taught Us "Never Give Up!"

One of my patients with dementia, a 76-year-old man, who was placed in a nursing home in Tajiri, exhibited sudden loss of consciousness with seizures in the evening of March 23 (12 days later the earthquake). Since his blood pressure started to decrease, I and another physician with my team requested to be transferred to the emergency hospital in the next town. The staff immediately carried him to the hospital by themselves, using the small amount of gasoline remaining; however, the hospital did not admit the patient due to the lack of capacity at that time. The situation did not allow us to call an emergency vehicle there and the lack of gasoline prevented us from taking him to any other hospital. The family told me by telephone that they had to resign themselves to the circumstance, and understand this seriousness of everybody who was in trouble, and asked me to do no more.

I understood the patient's family, but I reconsidered that he would recover after an adequate treatment. This patient was also a disaster victim! I thought, since this would have been a routine situation before the disaster, there should be no problem for admission. At midnight I called and awakened a pharmacist I knew personally and asked him to prepare the medication and drip infusion set. Meanwhile a staff member at the nursing home had found a gas station that would prioritize providing a welfare institution, and had waited all night long to get gasoline!

I stayed throughout the night and early the next morning, when the patient was in a very serious condition, a surgeon, who was my friend, found a bed for my patient at another hospital. The staff member, who obtained the gasoline early in the morning, brought the patient with the medication, and the drip infusion set, which arrived at virtually the same time (delivered by the pharmacy staff by bicycle), and finally the treatment was begun. The next day his general condition improved and his recovery was successful. This case taught us "Never give up!"

Effect on Dementia in Group Home

Group homes-A and B, which were run by a private company, were located in the coastal area of Kesen-numa city and were damaged by the disaster. Fourteen of the 17 elderly residents living in Group homes-A and B prior to the disaster were evacuated to another Group home-C, which was run by the same company. The number of residents in Group home-C prior to the disaster was nine, but almost tripled to 23 after the disaster, and several residents had to live together in one room. The compatibility of the residents living in the same room was investigated. Aggravation of dementia caused by overcrowding was anticipated, but surprisingly the behavior was improved, as in the following case.

A 90-year-old woman had severe dementia, with a MMSE score of 4. Before the earthquake, she did not eat her meals three times a day, but changed to eating three times a day, thereafter. She started to wipe the washed dishes dry, spontaneously. This situation was initially broadcasted in TV: “Unexpected changes in people with dementia” on April 27. “Improved eating,” “helping to dry dishes," all showed “improved motivation for living.” We investigated the changes, considering that psychological stress should be avoided, through careful interviews with the staff.

Twenty-three residents (mean age, 81.7 years) were studied for the diagnosis of dementing diseases, Mini-Mental State Examination, memories of the disaster, and the changes in the amount of care in daily life. They were all diagnosed with dementia according to DSM-IV. Dementing diseases were diagnosed by the NINCDS-ADRDA (McKhann et al., 1984) for Alzheimer’s disease, NINDS-AIREN (Roman et al., 1993) for vascular dementia, and both for mixed dementia, supported by the ischemic score (Hachinski et al., 1975). The severity of dementia was determined using the MMSE score: >20, mild; 10-19, moderate; and <9, severe. There were 14 patients with Alzheimer’s disease, six with mixed dementia, and three with vascular dementia, comprising three mild, nine moderate, and 11 severe dementia patients.

Memories of the disaster, including the earthquake, subsequent tsunami and the evacuation, were obtained in interviews. The previous tsunami was the Great Chilean (Valdivia) Earthquake in 1960 (51 years before the Great East Japan Earthquake 2011).

In contrast to the older refugees in other secondary areas, I found surprisingly that they had good judgments based on “episodic memory” and “knowledge” of the previous tsunami. They vividly described to me the situation of the previous tsunami as their experiences, not so far from the recent days. Regarding behavior, they responded with “Help each other” and “Escape” rather than “Stay at home,” according to their episodic and emotional memories. Figure 12 illustrates their memory of the disaster, as classified by dementia severity. Even people with severe stage dementia, about 25% of them recalled the previous Tsunami, and 50% of them recalled the Great East Japan Earthquake 2011.

Based on the behavioral neurological point of view, they actually had experienced the previous tsunami, and responded correctly, based on their episodic and emotional memories. This function is associated with the limbic structures of the temporal lobe, such as the amygdala and hippocampus, as illustrated in Figure 13. Patients with dementia, especially Alzheimer’s disease, had atrophied limbic structures, thus the correct response rate was decreased.

Two observers independently evaluated the changes in the amount of care in daily life required by the patients while eating, participating in recreational events, bathing, etc., and the relationships with other residents, in terms of consideration of others’ feelings, etc. The observers were from the managing staff of the company who had known all the residents for at least six months. The pre- and post-disaster assessments were performed one week prior to and one month after the disaster. The residents’ behaviors were assessed routinely during the first week of the month for one month. The disaster occurred on March 11, 2011, which was one week after the assessment had been done during the first week of March. Behaviors were defined as “improved” if both evaluators indicated the finding as “unchanged,” and the evaluations were defined as inconsistent and "aggravated" if both indicated as such, based upon direct observation. There were no changes in drug administration throughout this observation period.

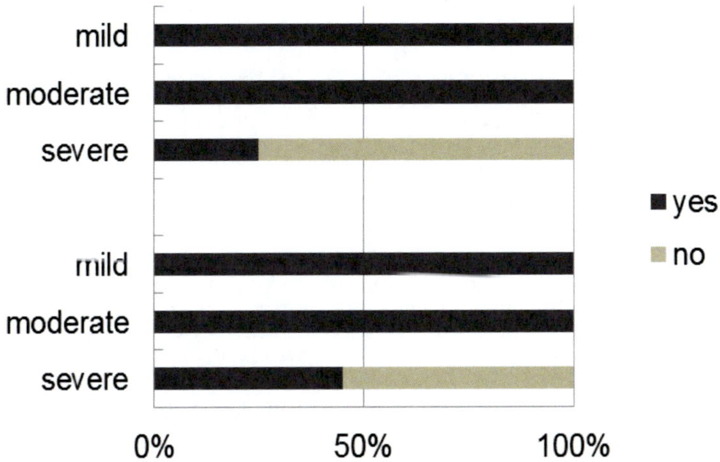

Figure 12. Memory on the disaster and dementia severity. Upper: The previous Tsunami. Lower: Great East Japan Earthquake 2011.

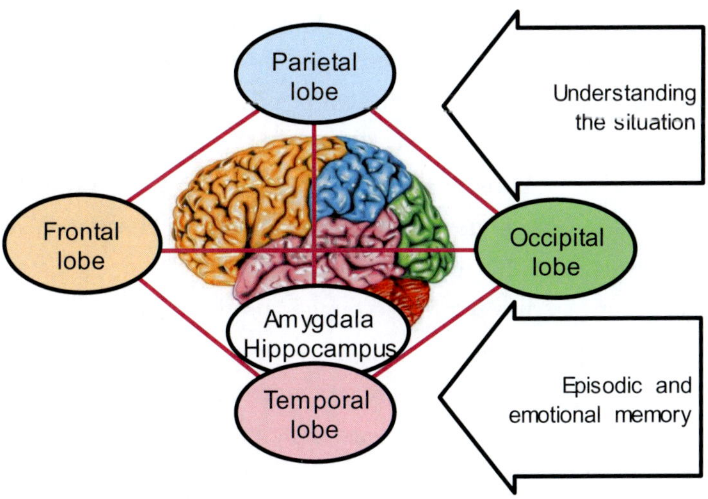

Figure 13. Brain functions associated with behavior under the disaster 2: episodic and emotional memory.

As Table 4 shows that, of the 23 residents, behavior "improved" in 12, remained "unchanged" in eight, and "aggravated" in three patients. The proportions of residents with "improved," "unchanged," and "aggravated" behavior were the same for residents with moderate dementia (3:3:3), but eight of the 11 residents with severe dementia showed "improved" behavior. Dementing diseases, severity, and behavioral changes among the residents in each room (rooms 1 to 10) are also shown in Table 4. The cases with "improved" behavior included "a mild vascular dementia resident who became more careful about TV volume in the dining room" (room 1), "a severe vascular dementia resident who began to eat more food," and "a moderate Alzheimer's disease resident who started to help with cleaning up after eating" (both in room 2). Some behaviors also "improved" in severe dementia residents, but no clear relationship with dementing disease was found.

Living in small groups and dining in large groups might have reminded the residents of their family and school lives. Residents with severe dementia appeared to have been particularly able to adjust to the new environment, likely due to stimulation of their remote memories. Their improved behaviors were probably due to activated function of the frontal lobe, especially the anterior cingulate, which was associated with "motivation," as illustrated in Figure 14, since similar areas have been reported to be activated through the reality orientation with reminiscence approach (Akanuma et al., 2011). Indeed, the most improved residents (11/13) had lived in households with three generations prior to their admission to the group home, and had also gone to school; however, two of the three residents with aggravated behavior were the oldest sisters from large families, and caring for their younger brothers had prevented them from going to school. All of the residents with mild to moderate dementia, but only half of the residents with severe dementia, retained the memory of the earthquake. Emotional memory was related to the amygdala (Schultz et al., 2009), and less retention of emotional memory was associated with greater atrophy of the amygdala.

Table 4. Conditions of patients and improvement or aggravation of behavior in each room

Room no.	Room type	Condition of residents			Improved behavior		Aggravated behavior
		Resident 1	Resident 2	Resident 3			
1	Room for 1 person	Mild VaD			Consideration of others in TV viewing		
2	Room for 2 persons	Severe VaD	Moderate AD		Eating	Cleaning up after eating	
3	Room for 2 persons	Moderate AD	Mild AD				
4	Room for 2 persons	Severe MxD	Mild MxD		Bathing and eating		
5	Room for 2 persons	Severe AD	Moderate MxD		Facial expressions		Motivation
6	Room for 3 persons	Severe AD	Severe AD	Moderate VaD	Sleeping and anxiety	Conversation	
7	Room for 3 persons	Severe AD	Moderate MxD	Moderate AD	Anxiety	Anxiety	Dis-orientation
8	Room for 3 persons	Severe AD	Severe AD	Moderate MxD	Drying washed dishes	Anxiety	Delusion
9	Room for 5 persons Japanese style	Severe AD Severe AD	Moderate MxD Moderate AD	Severe AD	Eating	Consideration of others in TV viewing	

AD = Alzheimer's disease, VaD = vascular dementia, MxD = Mixed dementia.

All of the residents with mild to moderate dementia, but only half of the residents with severe dementia retained the memory of the earthquake. Emotional memory was related to the amygdala (Schultz et al., 2009), and less retention of emotional memory was associated with greater atrophy of the amygdala.

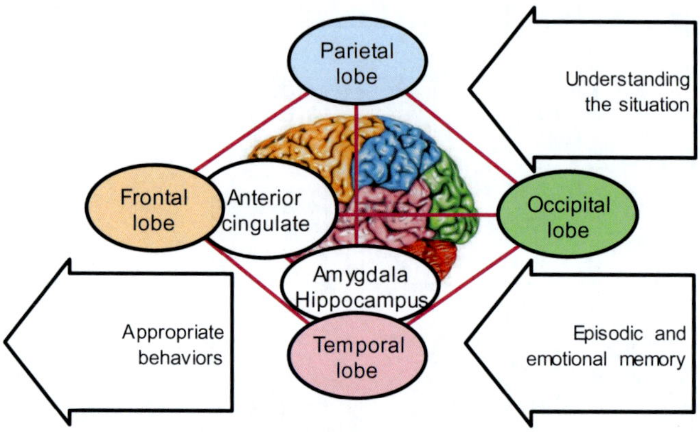

Figure 14. Brain functions associated with behavior under the disaster 3: appropriate behaviors.

Lessons Learned

- The lack of heating or quilts led to reductions in the patients' levels of activities of daily living (ADL). However, according to the natural rhythm based care, the sleep/wake rhythm cycle improved in some patients.
- The lack of gasoline was critical for consulting physicians, and the report of the 76-year-old man taught us to "never give up!"
- Older residents in the Group home had good memories of the previous tsunami, and exhibited adequate behaviors. A report on the 90-year-old woman who showed improvement in her behaviors suggested the positive effect of the group reminiscence approach. Living in small groups and dining in large groups might have reminded the residents of their family and school lives.

Chapter 4

Predicting Weakness Due to Disaster in the Community

The ability of humans to correctly judge a situation, to understand information, and to select an appropriate behavioral response is critical, especially for risk management after a disaster. Any kind of disaster-prevention broadcast or television news becomes meaningless unless the contents are correctly understood and processed by residents, allowing them to behave appropriately.

A number of community-based studies on aging and dementia have been performed, and the cognitive features of mild cognitive impairment (MCI) and dementia have been well reported with respect to memory impairment or executive dysfunction. However, the behavior of such subjects in response to a disaster has not previously been examined in detail. A kind of risk cognition task, rather than memory or executive function tests, could be useful to screen for weaknesses related to disaster in the community. As described in the Introduction, I believe that approaches to "brain-based disaster medicine" and "disaster-associated dementia medicine" have not been fully integrated or even investigated.

Herein, we focused on reconstruction support after the disaster from an academic perspective.

We hypothesized that the MCI subjects did not fully understand the content of television news with respect to the earthquake. We also hypothesized that that the MCI subjects would exhibit lower scores than healthy people on a risk cognition task, which would help explain the manifestation of inappropriate behavior in response to a disaster.

Summary of the Tome Project

As described earlier, after the disaster, we investigated the safety of residents in the affected communities. Most of the people requiring help were elderly and had previously been assessed as Clinical Dementia Rating (CDR) 0.5 (mild cognitive impairment: MCI). We examined how well they understood the TV news and whether they could make appropriate decisions. This community-based study of dementia and difficulties following a disaster started in Tome, northern Japan (Akanuma et al., 2016). The subjects were 188 randomly-selected older residents, who underwent CDR assessment (Hughes et al., 1982; Morris et al., 1993; Meguro. 2004), blood tests, MRI, and cognitive tests, including the original Visual Risk Cognition Task.

They were shown NHK broadcasts from the day to determine whether they could understand the content. Audio recordings of TV broadcasts of the NHK news from March 11, 2011 were presented to the subjects. This broadcast had the following content: "Here is the news for residents living in the area where a strong earthquake has occurred. Please act calmly. Switch off appliances and watch for fires after the quake has finished. First, protect yourself from falling or dropping objects. Currently, our studio in Shibuya, Tokyo is shaking. There is a danger of building collapses, landslides, and landslips. Protect yourself from falling or dropping objects. Protect yourself from falling or dropping objects. Remain calm and stay in a secure location for the time being, until the quake has finished. After the quake has finished, switch off appliances and watch for fires." The subjects were then asked three questions about the content:

Q1: Understanding the content. Question: What was the news about? Please tell me in as much detail as possible.

Q2: Understanding the instructions for behavior. Question: First, what should I do? Example of a correct answer: Protect yourself from falling or dropping objects, stay in a secure location for the time being, and calm down.

Q3: Understanding the future risk: Question: What did they say about the risk of possible danger (threats)?

Surprisingly neither the CDR 0 (healthy) nor the CDR 0.5 (MCI) subjects fully understood the TV news. Some subjects did not recognize the danger of aftershocks and engaged in risky behavior. If the elderly are unable to understand broadcasts on disaster prevention, these broadcasts lose value. The news in Japan may be too difficult to understand. From the behavioral neurological point of view, I propose that News broadcasts should use many pictures and/or photos and present the news at a slower pace, which would be consistent with the capacity of "older brains" to process information.

People with CDR 0.5, who exhibited the inappropriate behavior of not anticipating an aftershock, scored lower on the Visual Risk Cognition Task (Akanuma et al., 2016).

"Get Ready" Pyramid for Earthquakes

The "Get Ready" Pyramid for earthquakes demonstrates as follows (The Alfred E. Alquist seismic safety commission, 2009).

1. Learn how to be ready: Everyone can learn how to get ready. The base of the pyramid includes learning what to do to stay safe during an earthquake, and learning about the other layers.
2. Plan and organize: Everyone can plan what to do when disaster strikes. This involves making plans for what to do before, during, and after a disaster.
3. Train and practice: Training provides skills that are critical during disasters. Practicing these skills and having disaster plans can make a difference.
4. Manage supplies and equipment: Once people know what kinds of supplies they need, they can determine what they already have on hand and what they need to get.
5. Secure building contents: Many of the things people can do to secure the contents of their buildings are free. Some things, like latches and brackets, are low cost.
6. Protect building structure: Protecting the structure of buildings is a more costly investment, and an important part of getting ready for many people.
7. Safeguard finances: At the top of the pyramid, investing in ways to safeguard finances is a more costly outlay. Not all will benefit, but it can make a great difference for some people.

Normalcy Bias

For the “Get Ready” Pyramid for earthquakes, the most important item is the first layer: Learn how to be ready. Figure 15 illustrates the brain functions associated with behavior during the disaster 4: education. From the behavioral neurological point of view, we have to fully engage our frontal lobes in order to think “outside the box.”

As described in Chapter 2, when three vehicles belong to my staff were registered as “emergency cars” and tried to help victims in the areas, we faced the Ministries and agencies’ powers in their areas of jurisdiction.

They went to the National Police Agency in order to register. The information regarding the lack of gasoline and which station we should go to in order to supply the cars was within the jurisdiction of Ministry of Transport, which meant they had to go to another office. The information on the highways was within the scope of the Ministry of Land, Infrastructure, Transport and Tourism, which meant they also visited yet another office.

As noted above, I would not attribute the problems we experienced to bureaucracy; however, a simple and urgent system to unify information and decision making processes during a risky condition should be considered.

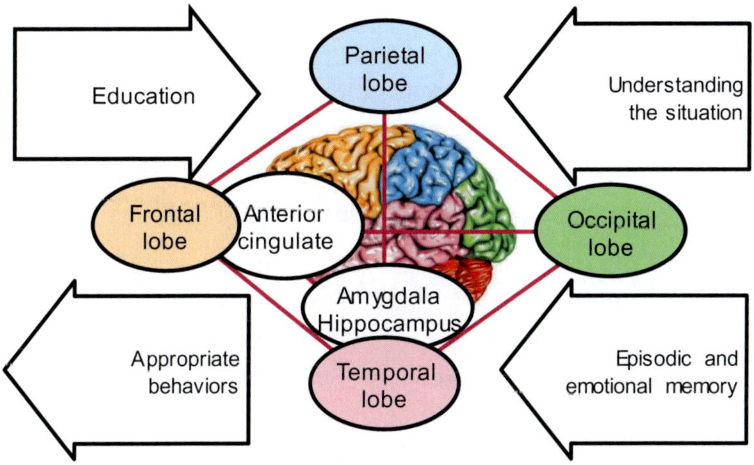

Figure 15. Brain functions associated with behavior under the disaster 4: education.

The term which became popular at that time was *soteigai* 「想定外」, which means "not presumed" or "beyond expectations." Indeed, people believed that the 1,000-year event would NOT happen tomorrow. A mega tsunami would NOT arrive, since few buildings collapse as a consequence of the earthquake itself. Also, a "super breakwater" would NOT be destroyed. In this sense, disaster medicine should be based on human psychology.

The normalcy bias, or normality bias, is a mental state people enter when facing a disaster. It causes people to underestimate both the possibility of a disaster and its possible effects. This may result in situations where people fail to adequately prepare and, on a larger scale, the failure of governments to include the populace in its disaster preparations.

The assumption that is made in the case of the normalcy bias is that since a disaster never has occurred, it never will occur. This can result in the inability of people to cope with a disaster when it does occur. People with a normalcy bias have difficulties reacting to something they have not experienced before, and tend to interpret warnings in the most optimistic way possible, seizing on any ambiguities to infer a less serious situation.

Figure 16 illustrates the Kamaishi City, guarded by the "Deepest Breakwater" (Guinness Records 2010), which was ultimately 63 m in height and 2 km in length, commenced construction in 1978 and was completed in 2008. Everybody was convinced that the breakwater could protect the people from a tsunami. However, it was destroyed and more than 1,000 people were killed (Modified from City of Kamaishi; Yomiuri Newspaper, April 2, 2011.).

Figure 17 illustrates the Taro Town, guarded by the "Super Breakwater, commenced construction in 1952 and was completed in 1982. Everybody thought that the breakwater could protect people from tsunami. However, it was also destroyed and more than 1,600 people were killed (Modified from Google Map; Sankei Newspaper, 2011, March 18).

A possible psychological explanation for the normalcy bias has been discussed. As shown in Figure 18, human judgment is comparative in nature. When people evaluate targets at all levels of complexity - ranging from simple psychophysical objects to complex social stimuli – they perform this in a comparative manner. Comparative evaluation forms a core mechanism in areas including person perception, decision making, etc. (Mussweiler. 2003). People who had not experienced the previous tsunami, considered that this earthquake was not such a serious problem, in a comparative manner, and had decided not to escape so urgently.

However, people who had experienced the previous tsunami, as described in Chapter 3, "Effect on dementia in Group Home (GH)," were able to escape with reference to their past episodic memories.

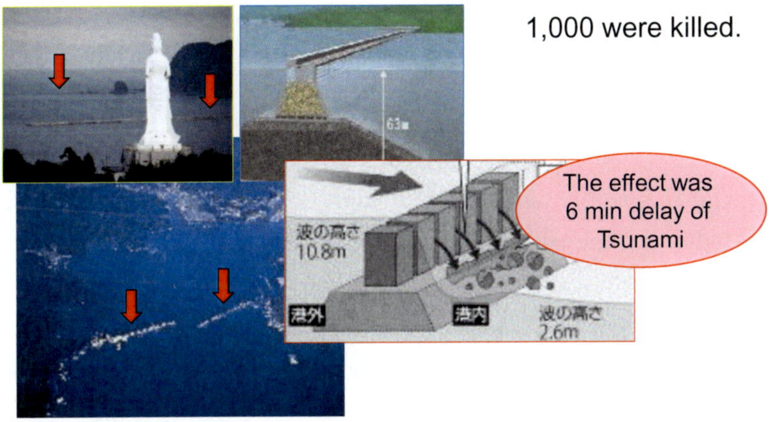

Figure 16. Kamaishi City guarded by the "Deepest Breakwater" (Guinness Records 2010). 63 m Ht for 2 km, commencement / completion: 1978 / 2008. *Modified from City of Kamaishi; Yomiuri Newspaper, April 2, 2011.*

Figure 17. Taro Town guarded by "Super Breakwater." 10 m Ht for 2.5 km, commencement / completion: 1952 / 1982. *Modified from Google Map; Sankei Newspaper, 2011, March 18. Photo: Kyodo News*

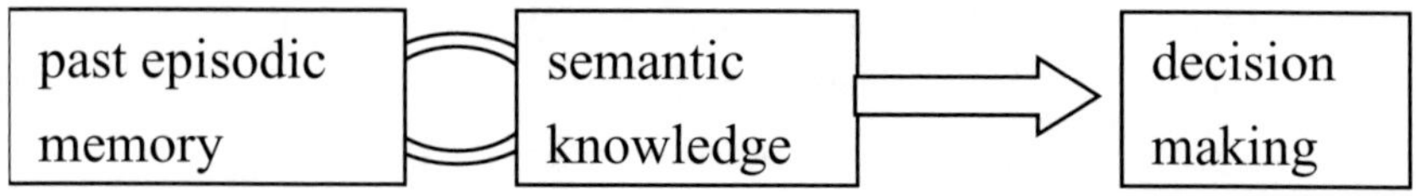

Figure 18. Comparison processes in social judgment: Mechanisms and consequences.

Lessons Learned

- Old-old people who behaved inappropriately during the Great East Japan Earthquake 2011 were vulnerable to the disaster. Risk cognition ability may be evaluated by using visually presented materials.
- It is noteworthy that even healthy subjects did not fully understand the content of the television news reports on the earthquake. If the elderly cannot understand the broadcasts on disaster prevention, these broadcasts lose value. The news in Japan may be too difficult to understand. News broadcasts using many pictures and/or photos and presentation at a slower pace are desirable, in order to be consistent with information processing capacity of elderly people.
- The normalcy bias, or normality bias, is a mental state people enter when facing a disaster. It causes people to underestimate both the possibility of a disaster and its possible effects.

Chapter 5

FY2012 Tome City Research and Examination Project for Dementia Countermeasures: Tome City Dementia Reality Check Project Report

As part of the FY2012 "Tome City Research and Examination Project for Dementia Countermeasures," this contracted research project was performed at the request of the Funded Course of Higher Cerebral Medicine for the Elderly, Tohoku University School of Medicine, in cooperation with Tome City.

The objectives of this research project were to estimate the prevalence of dementia in Tome City with its growing population of elderly residents, and to establish the basis for collaboration between the healthcare, welfare, medicine, and nursing care fields, in order to promote regional measures for dementia. Another objective was to provide opportunities for the enlightenment of dementia awareness among the local residents through research.

In addition, we also aimed to investigate the capability for coping behavior during a disaster among local elderly residents, as the region had experienced the Great East Japan Earthquake, and to prepare materials to provide more effective disaster countermeasures at medical/welfare sites. This report presents the analysis and results of the research performed in 2012.

Subjects and Methods

1. Target Area

Three administrative districts in Minamikata-machi, Tome City (Matsuba, Shintakaishi, and Hara)

2. Subjects

The subjects were 98 elderly residents (male: 35, female: 63) aged 75 years or older in the above-mentioned 3 administrative districts. From the Shintakaishi district, the subjects were selected randomly.

3. Methods

After obtaining written informed consent, we administered interviews and a questionnaire survey to residents aged 75 years or older in the above-mentioned districts, with respect to the health checkup on physical and cognitive functions and disasters. The data was subsequently analyzed cross-sectionally. We performed the following examinations:

1. General medical checkup (height, weight, blood pressure, pulse, and chest auscultation)

2. Neurological examination
3. Neuropsychological examination - including a test to reflect general cognitive function (MMSE), etc.
4. Questionnaire survey on the disaster - actions, knowledge, etc. with respect to the disaster
5. WHO-CIDI - to investigate past disease histories, such as depression and anxiety
6. Head MRI - to confirm the development of brain atrophy and cerebral stroke
7. Blood test - biochemical test, hematological test, thyroid function, and levels of B-complex vitamins
8. Questionnaire survey for family members of the subjects regarding daily life actions, past lifestyle habits, and actions during the disaster
9. Visits by a health nurse to conduct the interview for the life history, daily life situations, etc.

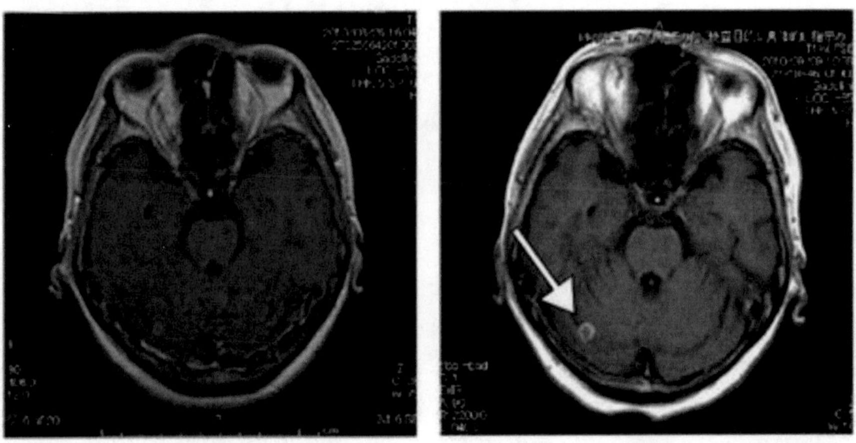

Figure 19. A 3-tesla (3-T) MRI device to acutely detect minute pathological changes. Quoted from: http://www.innervision.co.jp/suite/toshiba/seminarreport/101201/index.html.

The Clinical Dementia Rating (CDR) (Hughes et al., 1982; Morris et al., 1993; Meguro. 2004) was determined based on the results of the visits by the health nurse for the interview, the questionnaire survey for the families of the subjects, and the physician examinations.

For the MRI examination, we used a 3-tesla (3-T) MRI device to acutely detect minute pathological changes as illustrated in Figure 19.

The figures show a metastatic brain tumor, and show that a 3-T MRI can represent the tumor more clearly than a 1.5-tesla (1.5-T) MRI. Since the aim of this research was to determine the effects of minute cerebral lesions on cognitive function, we used the high-performance MRI device.

ANALYSIS

1. Dementia

The CDR was determined based on the results of the interview visits by the health nurse, a questionnaire survey for the families of the subjects, and the data obtained from the physician examinations. In addition, we confirmed dementia treatment for the subjects with CDR of 1 or higher (dementia).

2. Cerebral Stroke

We examined the relationship between a history of cerebral stroke and the MRI findings of cerebral infarction, the history of vascular risk factors (hypertension, diabetes, and hyperlipidemia), and the circumstances investigated in this research.

3. Depression

The depression state was assessed using the short version of the Geriatric Depression Scale (GDS). The short version GDS goes up to 15 points, with higher points indicating more severe depression. In this research, a score of 5 or higher points was evaluated as a "depressive state," in the screening of mild depression.

4. Disaster

Based on the results of the questionnaire survey for families of the subjects, we analyzed the situations of the use of evacuation centers immediately after the Great East Japan Earthquake 2011, the actions of the subjects after the earthquake had quieted down, poor health after the disaster, etc. In addition, based on the results of the questionnaire survey for the subjects, we analyzed the location and objectives of the evacuation centers, the actions taken when the subjects felt the destructive earthquake, and the actions in response to the lack of water and food.

RESULTS

1. Rate of Agreement

Figure 20 illustrates the rate of agreement. The overall rate of agreement among the subjects was approx. 50%, with no sex difference. The mean age of the subjects was 79.5 years (male: 78.4, female: 80.2). The rate of agreement in the Hara district (12/27) was lower than that in the Matsuba (19/33) and Shintakaishi (19/38) districts.

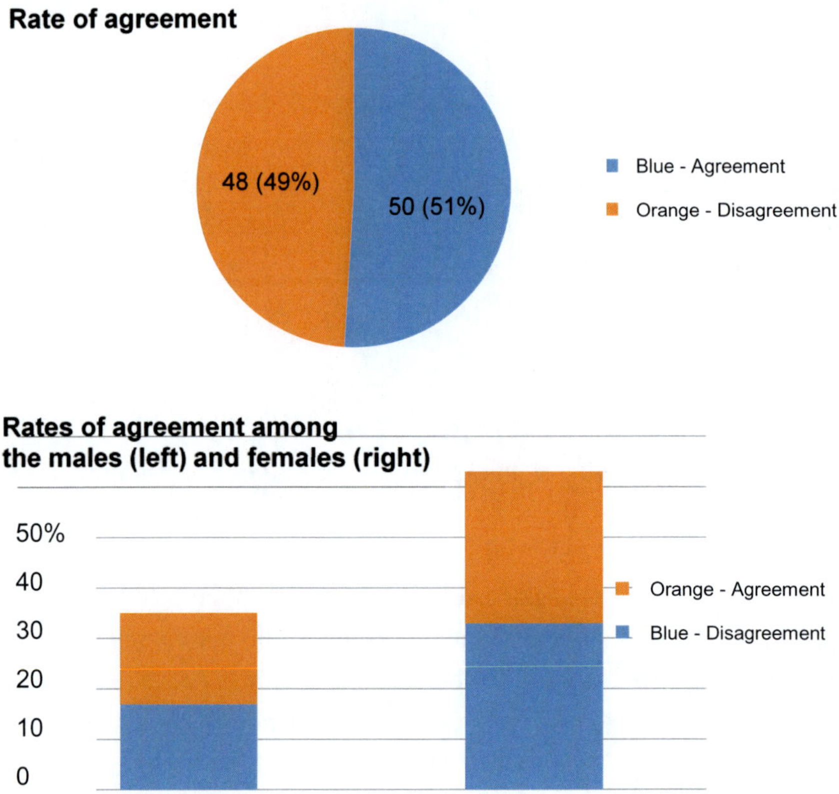

Figure 20. Rate of agreement.

2. Characteristics

Family Members Living with a Subject

As Figure 21 shows, two-generation families, including the elderly and their children, accounted for approx. 50%. Big families comprised of three generations were observed in about one-third of the families. Elderly couples and single-living elderly persons, for whom nursing care could be insufficient, accounted for approx. 20%.

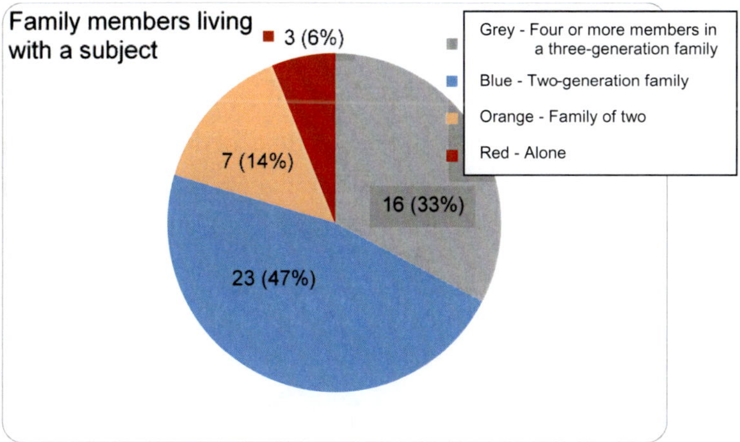

Figure 21. Family members living with a subject.

Duration of Education

As Figure 22 shows, the mean duration of education among all subjects was 9.6 years. The duration of 8 or 9 years was considered to be the duration required to graduate from the advanced course of ordinary elementary school or junior high school in the new education system. One-third of the elderly had 10 years or more of education.

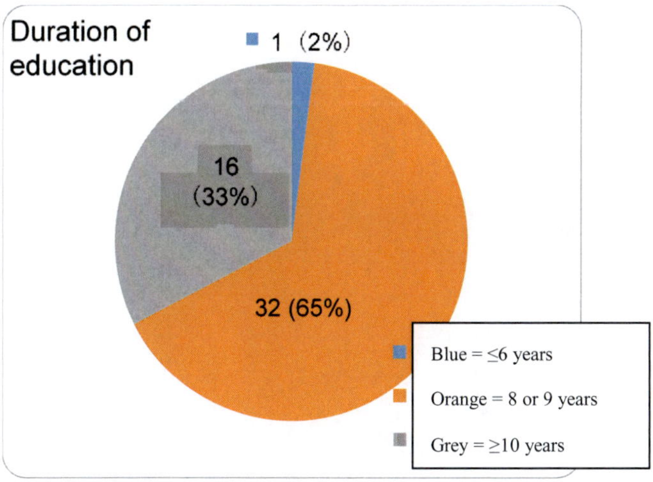

Figure 22. Duration of education.

3. Results of CDR Determination

As Figure 23 shows, among the subjects, 16% were determined to have dementia. The elderly on the borderline of healthy and dementia accounted for almost 50%, which was the highest level. These data were almost the same as the results of our investigation in other regions.

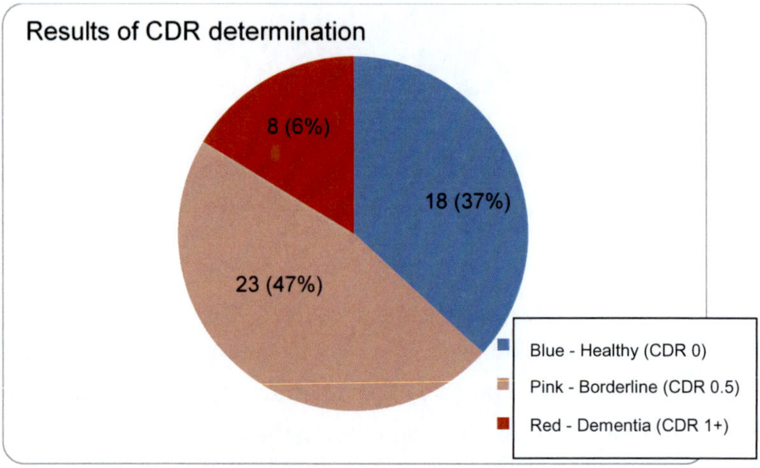

Figure 23. Results of CDR determination.

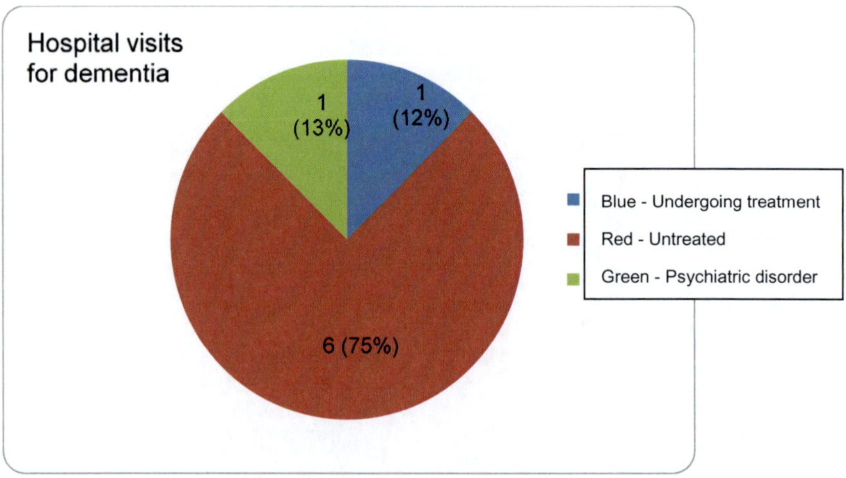

Figure 24. Hospital visits for dementia.

As Figure 24 shows, when we confirmed the visits to the outpatient department for dementia by the subjects with a CDR of 1 or more (dementia), only 1 subject made such visits. Six subjects were not diagnosed as having dementia, and did not receive any treatment for the disease. The remaining 1 subject had a psychiatric disease, and the reduction of the living ability by the psychiatric disease might affect the CDR determination.

4. Cerebral Stroke

History of Cerebral Strokc

As Figure 25 shows, 12% of the subjects answered they had a history of cerebral stroke.

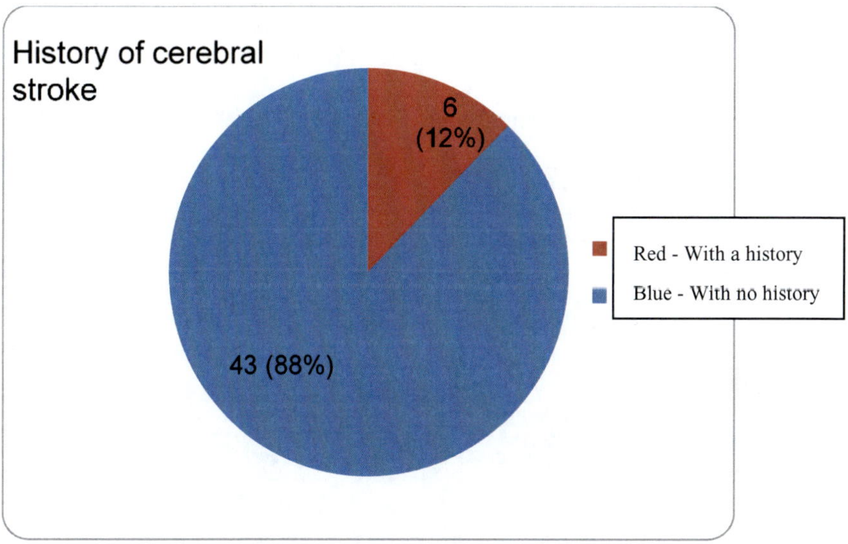

Figure 25. History of cerebral stroke.

As Figure 26 shows, among the 43 subjects with no history of cerebral stroke, cerebral infarction was found in 40% by MRI.

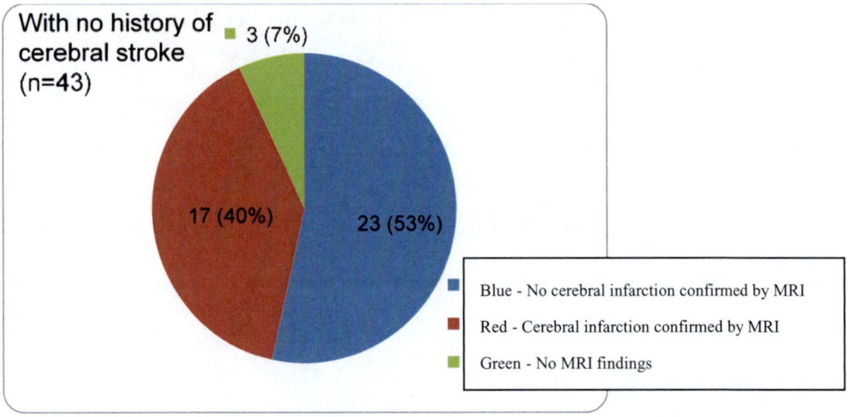

Figure 26. With no history of cerebral stroke.

Situations of Vascular Risk Factors

As Figure 27 shows, 60% of the subjects had hypertension, and the prevalence of diabetes or hyperlipidemia was approx. 20%.

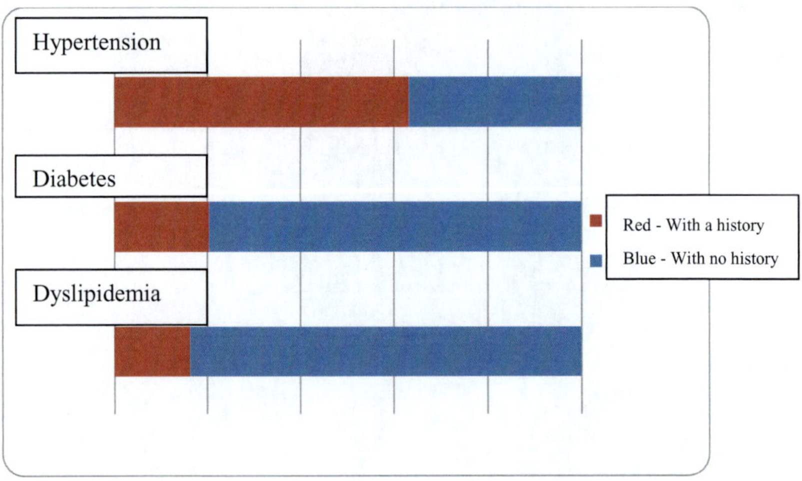

Figure 27. Situations of vascular risk factors.

As Figure 28 shows, among the 18 subjects who answered that they had no history of hypertension, almost 40% met the diagnostic criteria for hypertension (systolic blood pressure: 140 mmHg or higher, or diastolic blood pressure: 90 mmHg or higher).

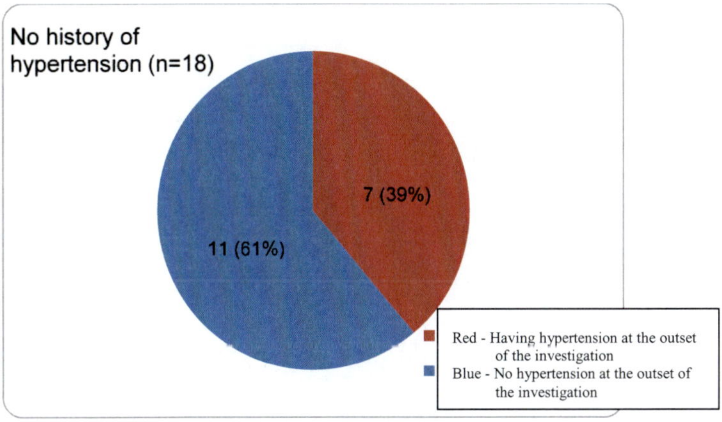

Figure 28. No history of hypertension.

As Figure 29 shows, among the 39 subjects who answered that they had no history of diabetes, 10% met the diagnostic criteria for diabetes (HbA1c ≥6.5%, or fasting blood glucose ≥126 mg/dl).

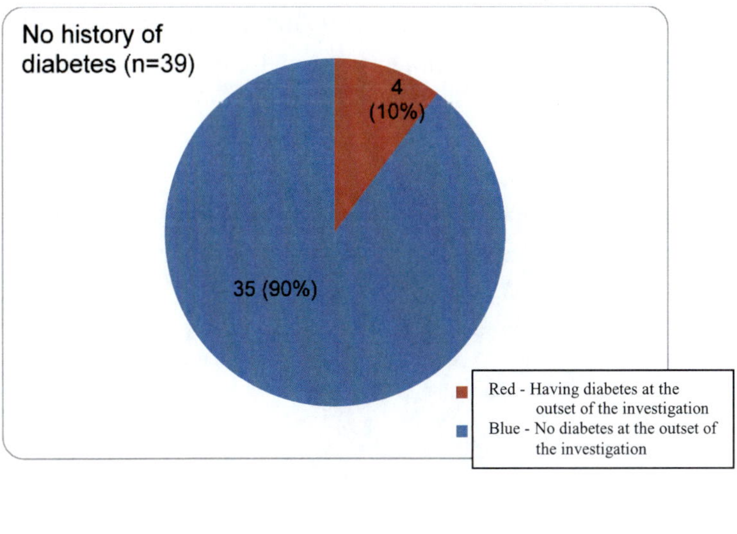

Figure 29. No history of diabetes.

As Figure 30 shows, among the 41 subjects who responded they had no history of hyperlipidemia, about one-third met the screening criteria for hyperlipidemia (LDL cholesterol ≥140 mg/dl, neutral fat ≥150 mg/dl, or HDL cholesterol <40 mg/dl).

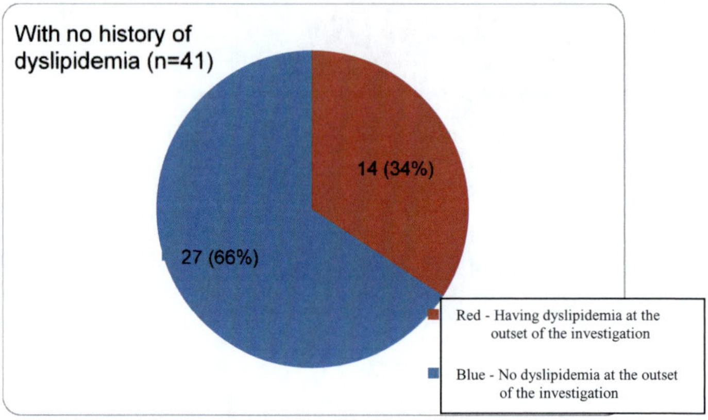

Figure 30. With no history of dyslipidemia.

Depressive State

As Figure 31 shows, depression was observed in approx. 40% of the subjects. The diagnosis of major depressive disorder was confirmed using the Mini-International Neuropsychiatric Interview (M.I.N.I.) during the hospital visits, but none of the subjects had major depressive disorder.

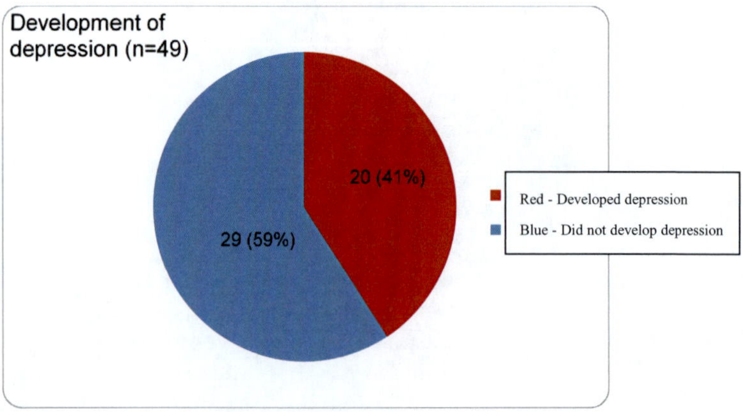

Figure 31. Development of depression.

5. Disaster-Related Items

From the questionnaire survey for families of the subjects

A) Use of Evacuation Centers

As Figure 32 shows, three subjects used evacuation centers following the Great East Japan Earthquake.

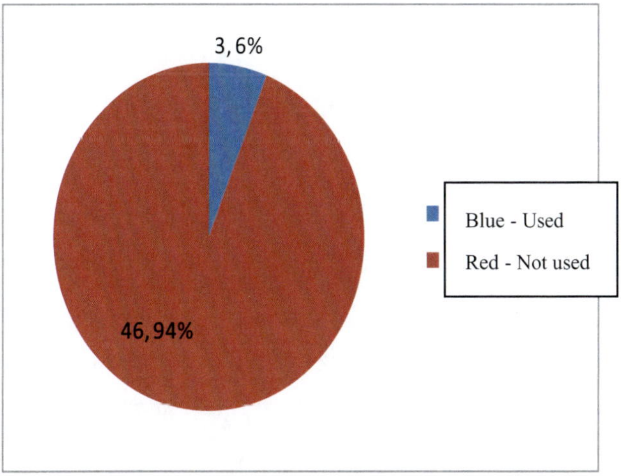

Figure 32. Use of evacuation centers.

As Figure 33 shows, the reasons included “anxiety about aftershocks,” “for relief,” “electric power outage,” etc., suggesting their anxiety at spending nights under the situation of continuous aftershocks.

Reasons for not using evacuation centers were shown in Figure 34. The major reasons for not using the evacuation center included no damage to the building and no problems with food.

“Other” reasons were:

- The grandchildren came to my house.
- I could live in cooperation with the neighbors (relatives).
- I had a car (large-size vehicle).

- I had no instructions.
- The damage was not so large as to require me to go to an evacuation center.

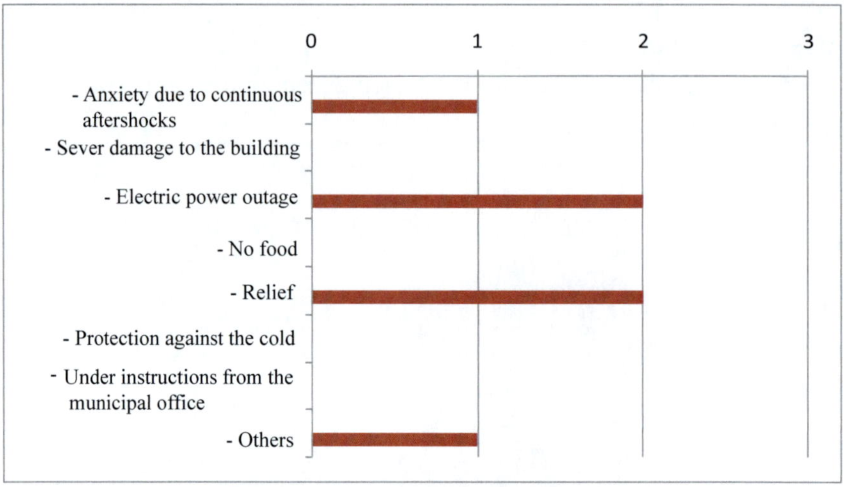

Figure 33. Reasons for the use of an evacuation center (multiple answers).

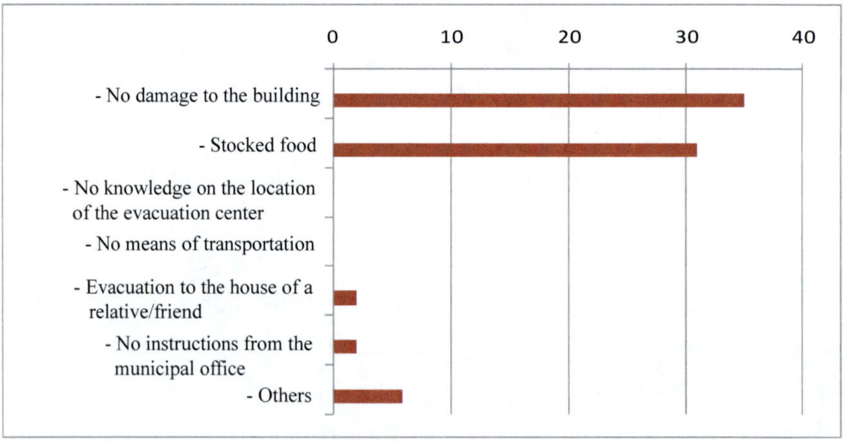

Figure 34. Reasons for not using evacuation centers (multiple answers).

As Figure 35 shows, the subjects who had known the location of the evacuation center beforehand accounted for 50%. 30% or more subjects somehow obtained information about the location of evacuation center.

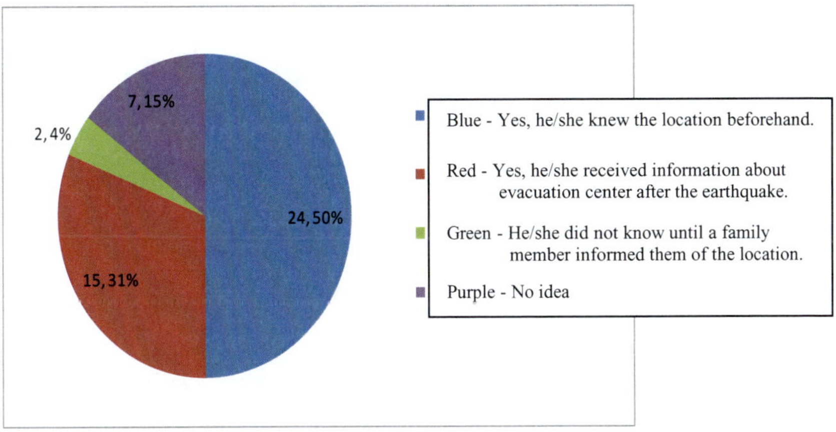

Figure 35. Did the subject know the location of an evacuation center?

B) Actions Taken Immediately after the Disaster

As Figure 36 shows, many subjects began to clear around the house or confirm the state of their houses with no regard towards aftershocks. This suggests the possibility that their knowledge with respect to the actions to be taken after a strong earthquake might have been insufficient before the disaster.

Others were:

- Prepared meals outdoors using a stove
- Stopped the car because the earthquake happened during driving.
- Received a call from my daughter.
- Opened the door.
- Visited the city hall to get water, and went to Sanuma to shop for food.

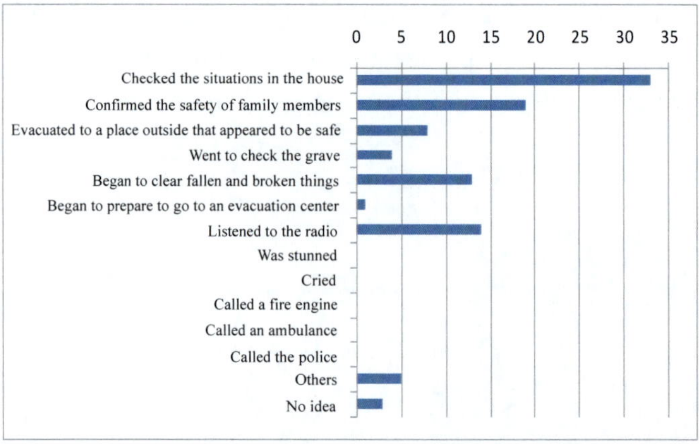

Figure 36. What actions did the subject take immediately after the earthquake quieted down?

C) Physical Conditions after the Disaster

None of the subjects were injured by the earthquake. In addition, only 1 subject showed poor physical conditions in the month after the disaster, due to "joint pain and common cold." None of the subjects suspended their normal medicine administration during the month after the disaster.

From the questionnaire survey for the subjects

D) About the Evacuation Center

As Figure 37 shows, more than 90% of the subjects had knowledge about when to use the evacuation center.

As Figure 38 shows, while almost all of the subjects seemed to understand the purposes of the evacuation center, some subjects showed insufficient understanding. We hope to examine the relationship between this and cognitive function in the future.

Others were:

- When a fire happens

- When an earthquake happens, with no idea about the concrete reasons
- When many troubles happen (with no idea about the concrete reasons), because it may be better to live together with other people
- Because the center is a place where administrative officers gather together
- Because the center is a place where young people gather together
- When I think I should go to the community center ← when a strong earthquake happens
- Because the center is a place for the activities of the fire prevention club, including a soup kitchen run by female members
- When a flood disaster or tsunami happens
- When experiencing flood disaster or radioactive damage
- Because the center is a place for the people who live in all-electric houses, where rice cooking cannot be performed with no electricity, and with no heating
- Because the center is a place for the people whose houses are damaged, and when the life lines are damaged.
- When floods happen

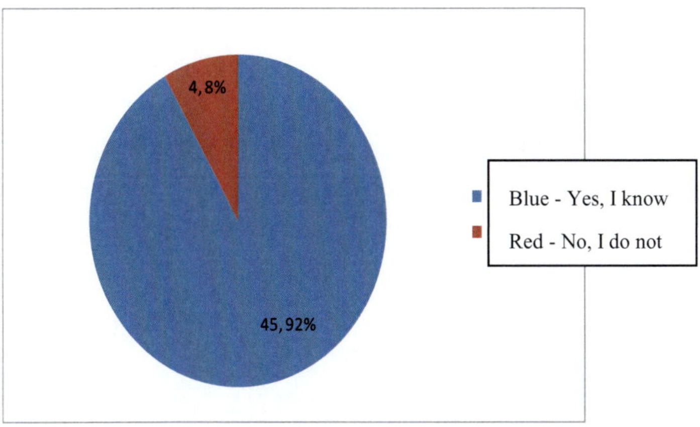

Figure 37. Do you know when an evacuation center should be used?

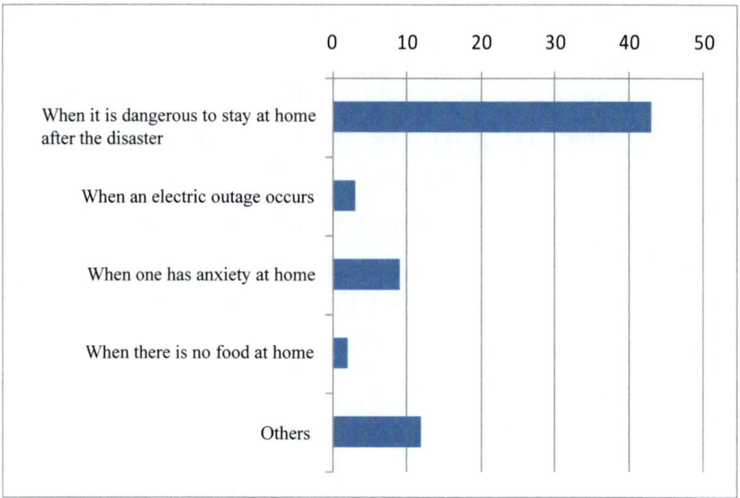

Figure 38. When should an evacuation center be used (When should you go to an evacuation center)?

As Figures 39 and 40 shows, the sites designated as evacuation centers in the Minamikata district are the Minamikata Agricultural Environment Improvement Center, Minamikata Elderly Welfare Center, Minamikata General Playground, Minamikata Elementary School, Minamikata Junior High School, Nishigo Elementary School, and Higashigo Elementary School, which was found to be understood by almost all of the subjects.

Others were:

- Having several experiences visiting the health center.
- Going to a building on a hill when a tsunami happens.
- Going to a building in Togo, maybe an Improvement Center (?), with no idea of any concrete information about the building.
- Going to the Improvement Center. (3 subjects)
- Going to a castle ruin on the hill in Hosokawa when a flood disaster happens.
- Going to Mt. Odake
- Perhaps going to a mountain behind my house, because it is easier if I have to evacuate.

- Going to a mountain when a tsunami happens, and in case of other types of disaster, going to a place where many people gather together.
- Not coming close to a building because it may be broken.

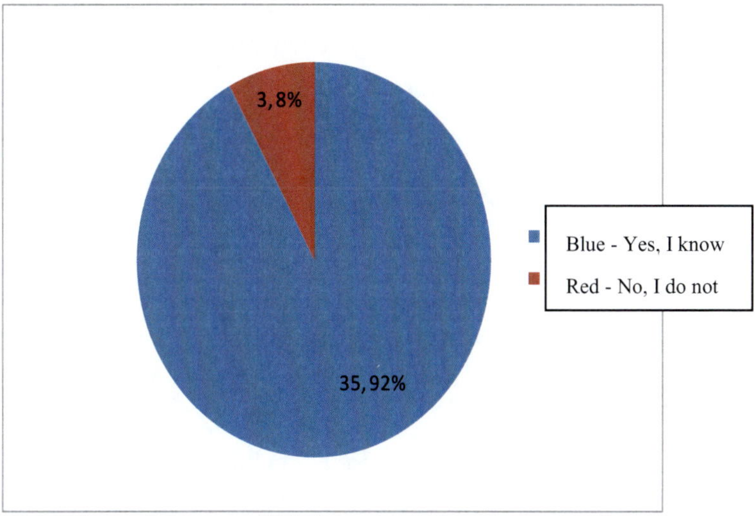

Figure 39. Do you know where the evacuation center is?

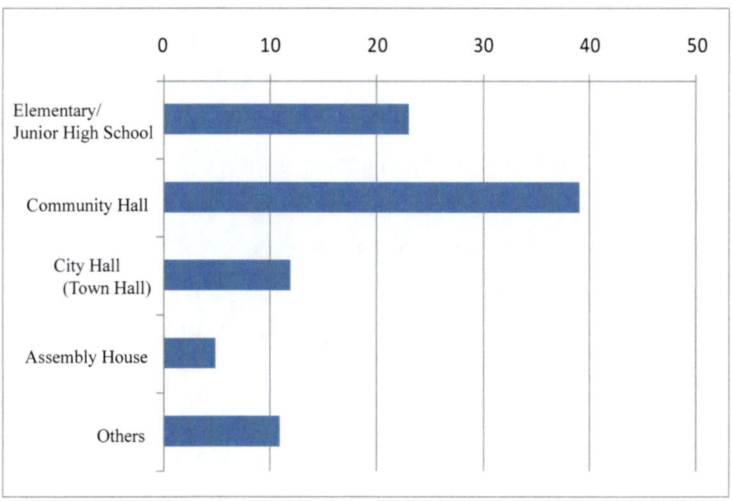

Figure 40. Most of the subjects knew the location of the evacuation center.

E) Actions Taken Following the Disaster

As Figure 41 shows, most subjects answered that they would take actions to avoid danger, such as "going outside," and "escaping to a place with no risk of falling objects," but some subjects answered that they would stay without moving. In addition, since some subjects answered that "they could not escape," and "they switched off the devices with fire," further examination will be required to understand the relationship with cognitive function.

Others were

- Switching off the devices with fire.
- Staying in place because I was familiar with the location. I stayed at home following the recent earthquake.
- Cannot escape due to difficulty in moving, and the outside is also dangerous.
- Protecting myself under the futon. (2 subjects)
- Holding on to a pillar.
- Addressing family members or neighbors.
- Addressing wife or mother (to escape).
- Escaping to a toilet because its pillars are strong, and follow directions from the family members when I am with them.
- Going to an evacuation center.
- Opening a door to secure an escape route.
- Making a judgment depending on the situation.
- Escaping with water and medications.
- Escaping with an emergency pack wrapped in a cloth.
- Getting into a place beneath the kotatsu. (2 subjects)

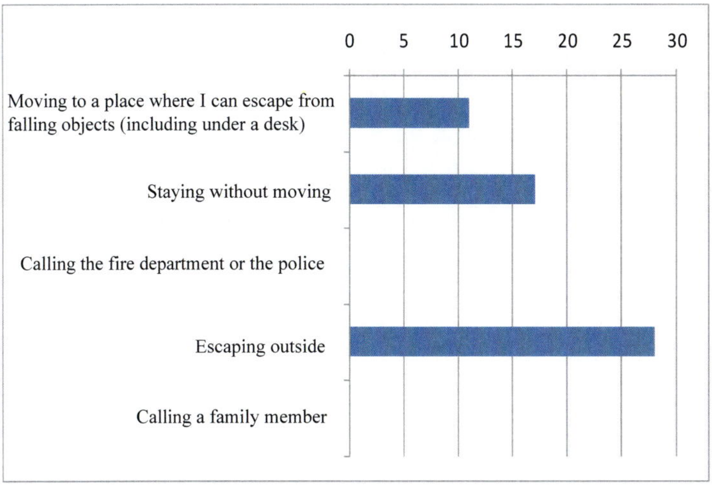

Figure 41. When you feel a strong earthquake at home, what do you do first?

As Figure 42 shows, many subjects considered they could supply their needs in cooperation with their relatives, which may be because many families were engaged in agriculture in some form. Under these situations, some subjects answered that "having never thought about it before," or "such a situation will not happen," and thus careful attention should be paid in future.

Others were:

- Never thought about it before. (8 subjects)
- Thought that such a situation will not happen. (2 subjects)
- Going somewhere with my rice cooker. (1 subject)
- Having no anxiety if my house is not damaged. (1 subject)
- Depending on family members (children, their spouses, and grandchildren) (20 subjects)
- Depending on relatives. (9 subjects)
- Using rations and a water tank truck (going to the city hall, town hall, community center, etc. to receive them). (13 subjects)
- Collecting information from broadcasting and publications. (3 subjects)

- Having no anxiety because we are a farm family with stocked rice, vegetables, miso, etc. (13 subjects)
- Going to stores to shop. (8 subjects)
- Having no anxiety because I can use my stockpile (stocks and reserved materials at home). (10 subjects)
- Using well water or water from a spring. (6 subjects)
- Consulting with the ward mayor. (1 subject)
- It remains to be seen. I have no idea. (3 subjects)

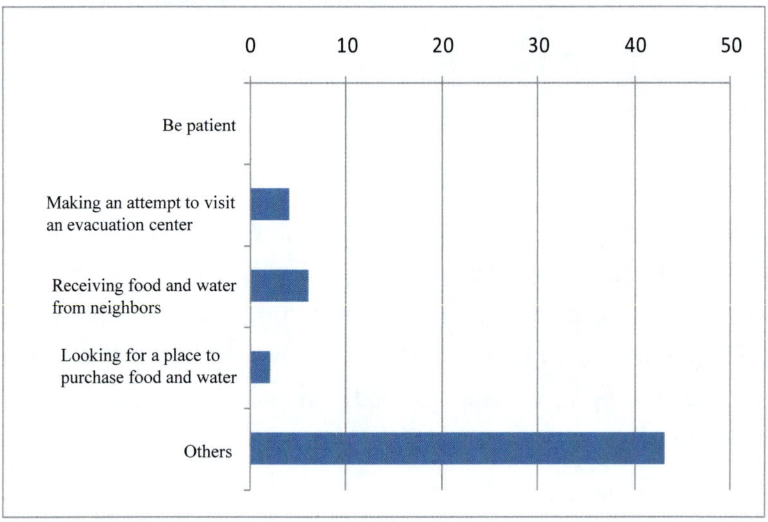

Figure 42. What do you do when water and food become insufficient due to the earthquake?

6. Relationship with the Neuropsychological Findings

Since the number of subjects was small in this investigation, a definitive conclusion cannot be reached. However, the following findings were clarified in a limited analysis.

We compared the responses to images with and without dangerous scenes, and the examination results for verbal memory tasks with and without danger information, to confirm their cognitive capability to understand disaster-related risks. We further examined their levels of understanding for TV news of NHK provided during the disaster.

Although the elderly with a CDR of 0.5 provided similar responses to the healthy elderly to normal scene images without danger information, their responses to the scene images with danger information were less concordant. Regarding verbal memory tasks, no differences were confirmed between the tasks with and without danger information. The understanding level for NHK news was low in both healthy elderly and the elderly with a CDR of 0.5.

Based on these findings, the cognitive capability of risks in the elderly with a CDR of 0.5 was considered to be low, which suggests the possibility that these elderly cannot respond appropriately to a disaster. In addition, the information provided by mass media when the disaster happened was difficult for the elderly to understand, and thus it may be necessary to provide information by using the community wireless system in a complementary manner. It will be possible to perform a more detailed investigation on the relationship between the cognitive capability for risks and the results of neuropsychological examination or head MRI findings by continuing the investigation with an increased number of subjects.

Discussion

1) Detection of Dementia and Medical Cooperation

The prevalence rates of dementia (CDR 1 or higher) and those on the borderline (CDR 0.5) were almost similar to those in the investigation of the residents of Kurihara City (12.5% and 50.0%, respectively). It was suggested that nearly 50% of the elderly residents in Tome City would have a CDR of 0.5. In addition, among the 7 subjects who were diagnosed as having dementia, 6 subjects received no diagnosis or treatment.

As a future task, a system should be established for the early detection/treatment of the elderly with a CDR of 0.5, who are included in the high-risk group of dementia patients. In addition, since the number of the elderly with untreated dementia may be high, educational activities for their families and in cooperation with home doctors will be required.

2) Prevention of Redevelopment of Cerebral Stroke

Among all subjects, 12% answered that they had a history of cerebral stroke. However, cerebral infarction was found in 40% of the subjects with no history of cerebral stroke by MRI. Since the possibility of redevelopment of cerebral infarction is high, cooperation with medical facilities is necessary.

Regarding the vascular risk factor, a history of hypertension was observed in many subjects. Even in the subjects with such a history, some subjects did not receive sufficient antihypertensive treatment. Hyperlipidemia was found in 34% of the subjects who had no history of the disease. In the future, it will be necessary to promote appropriate treatment for such risk factors, in order to prevent primary development or redevelopment of cerebral stroke.

3) Depression

The tendency towards depression was confirmed in 40% of the subjects. In the complete survey in Tajiri-cho, it was found that the tendency towards depression would be increased in residents who lived with more family members. In this investigation, the high prevalence of depression was likely to be affected by the living environment. Since it was suggested that the risk of dementia development might be increased when depression developed as a complication, aggressive intervention should be performed in individual regions.

4) Actions during or after the Disaster

Fortunately, there were no subjects who experienced trauma or damage to their health in the Great East Japan Earthquake. However, some subjects took inappropriate actions immediately after the disaster with no consideration of the possibility of aftershocks. In addition, the subject interviews suggested that a large number of the subjects could not take risk-avoiding behaviors.

Although the relationship with cognitive function should be examined as a future task, it will be necessary to provide simple and easy-to-understand information about the actions to be taken upon a disaster on a routine basis.

5) Future Tasks

The agreement rate was 50% in this research. However, since the persons with disagreement were likely to have dementia or cerebral stroke, a door-to-door survey, etc. should be conducted in the future. For this, cooperation with health nurses and care managers should be promoted. Furthermore, active measures are desired, especially in the Hara district, which had a low agreement rate.

In addition, since untreated dementia was found in many subjects, these subjects should be treated appropriately. It is planned to provide this information to their home doctors, and to recommend they visit the outpatient department of forgetfulness at the Tome Citizen Hospital. Since untreated dementia is likely to be found, even in other districts, it will be important to continue similar investigations, establish a cooperative system for medical nursing care and welfare in individual regions, and promote countermeasures for disadvantaged persons.

Chapter 6

ETHICAL PROBLEM

THE "THREE PRINCIPLES OF EVACUATION" PROPOSED BY PROFESSOR KATADA

According to Prof. Katada (Katada T. 2012), who provides education for tsunami disaster prevention, mainly in Kamaishi City, Iwate Prefecture, education for disaster prevention includes "education with threat," "education of knowledge," and "education for attitude."

Regarding "education with threat," careful attention is required because people may learn to dislike areas that are prone to frequent disasters. Although human brains have the ability to forget scary things in order to maintain their mental balance, it is sometimes necessary to deal with "great behaviors of nature" to approach nature and continuously obtain its blessing. Therefore, the "wisdom to make it through a disaster" is essential.

On the other hand, regarding "education of knowledge," a hazard map prepared by the local government is a typical example, but this may induce the fixation of a disaster image. Based on the data showing that the survival rate of elementary and junior high school students in Kamaishi City was 99.8% (100% under school control), it was suggested that "education for attitude" would be important, and thus "three principles of evacuation" were proposed. The principles are: 1) "Do not get caught up in an assumption," 2) "Do the best under the situation," and 3) "Be a refugee who takes the initiative to evacuate."

With respect to the first principle, 1), firstly, a hazard map, which is prepared assuming flood damage by a tsunami, is just an assumption made by humans based on the past data from tsunamis. We have to face the marvels of nature. In Japan, the Disaster Countermeasure Basic Act came into force in 1961, and disaster prevention measures have been promoted by local governments. Although a "100-year disaster" represents a good achievement of disaster prevention measures, residents may be overprotected from disaster because they come to believe that "prevention measures should be taken by the government." For example, the local government made the Guinness Super Breakwater record in the city. However, quite a few people died due to their decreased anti-disaster consciousness, since they had put too much confidence in the breakwater, and they did not evacuate due to the lack of an evacuation advisory from the government. Such cases were caused by "excessive belief about the assumption," not by an "unexpected situation," or an "inadequate assumption."

Next, with respect to the second principle, 2), it is necessary for individuals to do their best to protect their own lives when a disaster occurs. The power of nature is sometimes greater than people can predict. It is important to evacuate based on one's own judgment. Finally, with respect to the third principle, 3), humans have a tendency to ignore danger information to themselves (normalcy bias). It is necessary to evacuate first without paying attention to the surrounding people. It is suggested that the lives of the surrounding people would be saved by protecting one's own life. Related to this, I would now like to discuss "tsunami-tendenko," a word that became famous after the great disaster.

TSUNAMI-TENDENKO

"Tsunami-tendenko" is a word from the Tohoku dialect, created by Mr. Yamashita in 1933 based on his own experience; it is not an old saying (Yamashita. 2008). The word means that people should evacuate themselves individually without paying attention to others, even their own family, when a tsunami occurs.

If family members evacuate individually, not as a unit, vigorous young people may survive, but children and old people may not. When taking into consideration the importance of mutual mental and physical support among family members, it is generally considered that evacuation together with other family members would be advantageous for survival. According to Dr. Kodama (2015), there are two critical points for "tsunami-tendenko." One is that the word is egoistic and contrary to ethics, and the other is that the word goes against the nature of humans to save others. Therefore, it is impossible or difficult to believe the concept of the word from a psychological perspective.

For the first point, Dr. Kodama counter argues that "tsunami-tendenko" is an action that could prevent circumstances in which family members fall together, using a famous ethical word of "prisoners' dilemma." Actually, there were quite a few cases in the Great East Japan Earthquake 2011, in which family members were hit together by a tsunami while the parents went back home to look for their children, or family members waited for other members so that they could evacuate together. Therefore, "tsunami-tendenko" may be justified, and considered not to be egoistic from the perspective of indirect "consequentialism," because the final number of survivors would be increased as a result.

Regarding the second point, "tsunami-tendenko" may be contrary to the "deontological maxim," which suggests that people should save dying people. However, when the Great East Japan Earthquake occurred, many people actually followed "tsunami-tendenko," and the Ministry of Education, Culture, Sports, Science and Technology (MEXT) is currently planning to teach the word to students. Thus, criticism for the idea may be limited. Dr. Kodama continued to argue that "Tsunami-tendenko" is the best idea when the safety of a person to be saved is unclear, but it is obvious that the person can evacuate by him/herself. However, when the safety is unclear, and it is obviously impossible for the person to evacuate by him/herself, it is psychologically difficult to believe in this approach. This is because people must give up the life of their family member if they follow the idea of "tsunami-tendenko." However, there is the possibility that a rescue team could save the family member, and thus the idea of "tsunami-tendenko" is considered to be correct. Dr. Kodama suggests that rescue specialists, such as police officers and firefighters, should not follow the idea of "tsunami-tendenko." Of course, their lives should not be sacrificed, but they are specialists who are able to rescue other people while minimizing the risks for themselves.

As a conclusion, it should be highlighted that "tsunami-tendenko," which is a simple thesis, should be taught as a method to maximize the number of survivors based on the relationship of mutual trust, and that rescue teams should act based on another code of conduct. As noted above, Prof. Katada also suggested that "tsunami-tendenko" might be a painful approach to prevent repeated trauma in the destruction of a family due to a family bond, based on the consistency of "a person to be saved" and "a person who saves others." These concepts were suggested in his education seminars for disaster prevention. In addition, he interpreted the real meaning of the idea as that individual family members should be responsible for their own lives, and that the family should have a relationship of mutual trust to save one's own life.

Problem of a Dual Code of Conduct

However, there should be further discussion on the code of conduct for rescue teams. Let me show another example of risk. "What actions would you take if you saw your neighbor's home catching fire?" The correct answer to this question, which was also introduced in this book, is

2 or more of the following answers: "extinguish," "inform," and "call a firehouse." The answer, "evacuation, expecting a rescue team to extinguish" is not included among the correct responses. As noted above, however, according to the textbook of behavioral neurology, "judgment" and "insight" are described as a set. There are cases in which a big fire disaster could be prevented by performing first-aid firefighting, as much as possible, but the physical capability should be taken into consideration in some cases. One of my outpatients at the memory clinic said, "I would perform firefighting when I was young, but now have no choice but to evacuate because I have no confidence in my physical strength." This simply demonstrates the idea of "fire-tendenko." Although this is a good answer based on self-observation of physical function, it is unfortunately the correct answer in a psychological test. There is another example. Although observation by local residents is important to prevent juvenile crimes, it is natural for local residents to only report an accident with no direct actions to the criminal with a weapon, even if such residents are instructed not to escape ("crime-tendenko"). This can also be seen in the observation of patients with dementia who show behavioral abnormalities by local residents.

This is the same for diseases, as "health risks." According to a diabetes specialist on our team, many patients with diabetes mellitus have a mental problem, a so-called "dependency on foods." They take diabetes drugs to allow them to eat sweets as they like. This is analogous to putting the cart before the horse. They have the idea of "disease-tendenko," and their drugs are a sort of "rescue team." Diabetes mellitus is a representative lifestyle disease, and the drug is just an aid. Unless the patient faces the disease in a proactive manner by controlling their own life, the treatment for the disease will not be effective. This is same as the "disaster prevention education for attitude" suggested by Prof. Katada above. However, when surgical treatment is necessary for a serious disease, there is no choice but to depend on a physician.

It is important to judge the balance between risk and one's own capability to deal with the problem. The types of response of society and individuals to disaster suggested by Dr. Hirose (2004) are shown in Table 5.

Table 5. The types of response of society and individuals to disaster

		Grade of disaster		
		Big	Small	Under threshold
Controllability of disaster	Yes	overactive response (panic)	Reaction based on cost-benefit judgment	ignore
	No	resignation	patience	

Modified from Hirose. Shueisha 2004.

In a great disaster, otherwise controllable situations, including evacuation, may lead to an overactive response (panic). However, it is also problematic that people become excessively dependent on public administration after they receive education recommending escape from all risks, based on an exaggerated interpretation of "tendenko."

Based on their own intent, individuals should face risk constructively and do their best, as much as possible. However, when people have to face a great disaster, they have no other choice but to evacuate first, and become dependent upon professionals for rescue. However, I feel strongly that people should always have the same spirit as the rescue teams, that is, "public attitude." Rescue team members are also the same as everyone else. According to the Bible, all humans are like a "deceitful bow." Precisely because every person has problems, a spirit of mutual cooperation is necessary.

According to the economist, Herbert Simon (1993), the "action of altruism" is activated under emergency rules. The "action of altruism" may not produce benefits to the persons who perform the action, but may increase the benefits to society as a whole. Unless the total loss for the people who perform the "action of altruism" is greater than the benefits to the whole of society, the action contributes to the maintenance of social functions. In addition, when persons who do not perform the "action of altruism" are punished, including ostracism from community, the "action of altruism" is promoted. Although the forced "action of altruism" is not acceptable, there would be no applicants for rescue teams, unless such a voluntary spirit is grown and regarded highly in society. This can also apply to police officers, firefighters, members of the Self-Defense Forces or Japan Coast Guard, and healthcare workers.

CONCLUSION: NATURE, HISTORY AND ASIA

SAFETY CULTURE: AN IMPORTANCE OF CONCEPT FOR COEXISTING WITH NATURE

The Great East Japan Earthquake 2011 caused serious damage to Japan, and more than 15,000 people were killed. Actually, three relatives in my family passed away. However, the disaster gave us a chance to reconsider many essential things. One of which is the importance of coexisting with nature.

Since about 7% of the volcanoes in the world exist in the islands of Japan, Japanese people have long been coexisting with earthquakes. Since the Great East Japan Earthquake 2011, the volcanoes in the islands of Japan are considered to have become more active. There are more than 110 volcanoes in Japan. Earthquake do not seem to benefit humans; however, volcanoes do provide us with a few merits; for example, there are more than 15,000 sources of hot springs in Japan. Volcanoes have provided the ground with many productive minerals. God, silver, and copper mines have been famous throughout history. In addition, typhoons, also "natural disaster" provide waters to the ground. Therefore, instead of being a dry desert, the islands have been fruitful, with about 300,000 people living in *Jomon* Era 4,500 years ago. Since then, the Japanese population has increased without marked immigration, and has established a unique civilization. Furthermore, people have learned that we should help each other during such disasters. At the time of the Great East Japan Earthquake 2011, TV abroad broadcasted that Japanese people surprisingly did not plunder the goods of others, and waited lines to receive relief supplies. All were due to the disaster. This is to say, "high risk, high education."

Catfish and Earthquake (Ukiyoe)

The second thing to reconsider is the importance of learning about history. As described in the Chapter 1, we should accept what we can learn from history, since "history repeats itself." An old document from the 9th century recorded a similar earthquake and tsunami, and an old shrine was built as a marker of the previous arrival point of the tsunami, which was almost the same location as the modern day tsunami of the Great East Japan Earthquake 2011.

Traditionally, the Japanese people have believed that earthquakes were due to hyperactivity of catfish. Figure 43 illustrated the Ukiyoe pictures in the *Edo* era (19th century). The first picture (upper left) shows that the catfish is punished by God, and the second picture (lower left) indicates that people are attacking the catfish as encouraged by God. The third picture (central) reveals that the catfish reflect upon their past activities, change their minds, and help the victims (restoration), and the fourth picture (right) shows that God gives money freely to the people (economic revitalization).

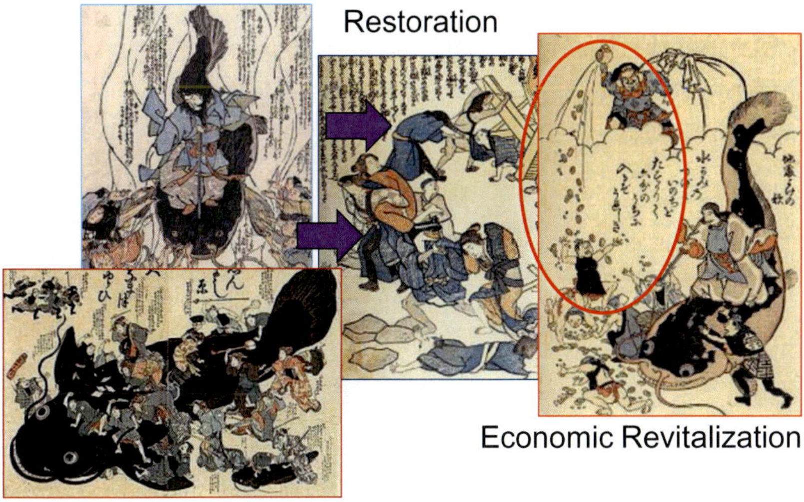

Figure 43. Catfish and earthquake (Ukiyoe). Modified from Namazu-e of Ukiyoe, Japan. 1855.

Anomalous electromagnetic signals were detected several hours before the occurrence of large earthquakes 150 km from the location of the electrodes (Fujinawa et al., 1990). So, is there any relation with catfish activities?

JAPAN, TAIWAN, AND SOUTH EAST ASIA: A NATURAL COMMUNITY WITH A COMMON DESTINY

Japan, Taiwan, and South East Asian countries face the same disasters because of the situation of the tectonic plates, as illustrated in Figure 44.

There is a graduate course in "Human Security" at my university. Most graduate students are from Indonesia. One student told me that the Indonesian government faces difficulties in preparing for disaster with a unified language. Indeed, more than 700 regional languages are spoken in Indonesia's numerous islands. However, disasters do not select a "language." However, I am certain that there are some lessons from their history that have not been recorded by popular language. We should establish a network based not only on local culture and language, but also a global proposal based on a common language.

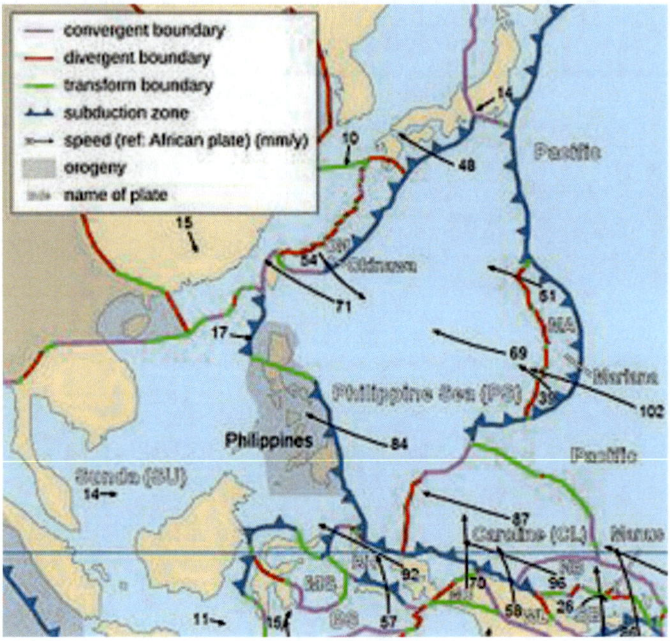

Figure 44. Japan and South East Asia: A natural community with a common destiny. *Philippine Plate map.*

References

Akanuma K, Meguro K, Meguro M, Sasaki E, Chiba K, Ishii H, Tanaka N. Improved social interaction and increased anterior cingulate metabolism after group reminiscence with reality orientation approach for vascular dementia. *Psychiatry Research: Neuroimaging* 2011;192(3):183-187.

Akanuma K, Kei Nakamura K, Meguro K, Chiba M, Sergio Ramón Gutiérrez Ubeda SR, Kumai K, Kato Y, Onuma J, Kasai M, Nakatsuka M, Seki T, Tomita H; The Tome Project Members. Disturbed social recognition and impaired risk judgment in older residents with MCI after the Great East Japan Earthquake 2011: *The Tome Project. Psychogeriatrics* 2016 [Epub].

Baldessarini RJ, Tondo L, Ghiani C, Lepri B. Illness risk following rapid versus gradual discontinuation of antidepressants. *American Journal of Psychiatry* 2010;167:934-941.

Folstein MF, Folstein SE, McHugh PR. "Mini-mental state." A practical method for grading the cognitive state of patients for the clinician. *Journal of Psychiatry Research* 1975;12:189-198.

Fujinawa Y, Takahashi K. Emission of electromagnetic radiation preceding the Ito seismic swarm of 1989. *Nature* 1990;347:376-378.

Fujiwara T, Sugawara M, Ookura Y, Mimune M. Nihon Sandai Jitsuroku, 892-901: Article 16 [Jpn]. (藤原時平、菅原道真、大蔵善行、三統理平. 日本三代実録892-901巻16).

Furukawa K, Ootsuki M, Kodama M, Arai H. Exacerbation of dementia after the earthquake and tsunami in Japan. *Journal of Neurology* 2012;259:1243.

Hachinski VC, Iliff LD, Zilhka E. Cerebral blood flow in dementia. *Archives of Neurology* 1975;32:632-637.

Hirose H, Why humans fail to escape? *Disaster Psychology* (Hito wa naze nigeokureru no ka? Saigai no shinrigaku). [Japanese] Shueisha, Tokyo, 2004. ISBN 978-4-08-720228-1 C0211.

Hughes CP, Berg L, Danziger WL, Coben LA, Martin RL. A new clinical scale for the staging of dementia. *Br J Psychiatry* 1982;140:566-572.

Ishii H, Meguro K, Yamaguchi S, Ishikawa H, Yamadori A. Prevalence and cognitive performances of vascular cognitive impairment no dementia in Japan: the Osaki-Tajiri Project. *European Journal of Neurology* 2007;14:609-616.

Ishii H, Ishikawa H, Tashiro M, Yamaguchi S, Meguro K. Decreased cortical glucose metabolism in converters from CDR 0.5 to Alzheimer's disease in a community: The Osaki-Tajiri Project. *International Psychogeriatrics* 2009;21:148-156.

Kasai M, Meguro K, Nakamura K, Nakatsuka M, Ouch Y, Tanaka N. Screening for very Mild Subcortical Vascular Dementia Patients Aged 75 and above using Montreal Cognitive Assessment and Mini Mental State Exam in a Community: The Kurihara Project. *Dementia and Geriatric Cognitive Disorders Extra* 2012;2:503-515. Epub 2012 Nov 10.

Katada T. *Disaster prevention to save lives* (Hito ga shinanai bosai) [Japanese]. Shueisha, Tokyo, 2012. ISBN 978-4-08-720633-3 C0236.

Kobayashi S, Hanagama M, Yamanda S, Yanai M. Home oxygen therapy during natural disasters: lessons from the great East Japan earthquake. *European Respiratory Journal* 2012;39:1047-1048.

Kobayashi Y, Takahashi Y, Seki T, Kaneta T, Amarume K, Kasai M, Meguro K; the Kurihara Project Members. Decreased Physical Activity associated with Executive Dysfunction Correlates with Cognitive Impairment among Older Adults in the Community: A Retrospective Analysis from the Kurihara Project: Implication for "Exercise" Intervention for Dementia Prevention. *Dementia and Geriatric Cognitive Disorders Extra* 2016;6:350-360.

Kodama S. Tsunami-tendenko and morality in disasters. *J Med Ethics* 2015;41:361-363.

Kumai K, Meguro K, Kasai M, Nakamura K, Nakatsuka M; the Kurihara Project members. Neuroepidemiologic and neurobehavioral characteristics of Motoric Cognitive Risk syndrome among the old-old population: The Kurihara Project. *Dementia and Geriatric Cognitive Disorders Extra* 2016;6:176-182.

McCurry J. Japan: the aftermath. *Lancet* 2011;377:1061-1062.

McKhann G, Drachman D, Folstein M, Katzman R, Proce D, Stadlan EM. Clinical diagnosis of Alzheimer's disease: report of the NINCDS–ADRDA Work Group under the auspices of Department of Health and Human Services Task Force on Alzheimer's disease. *Neurology* 1984;34:939-944.

Meguro K, Ishii H, Yamaguchi S, Ishizaki J, Shimada M, Sato M, Hashimoto R, Shimada Y, Meguro M, Yamadori A, Sekita Y. Prevalence of dementia and dementing diseases in Japan: the Tajiri Project. *Archives of Neurology* 2002;59:1109-1114.

Meguro K. Ishii H, Yamaguchi S, Ishizaki J, Sato M, Hashimoto R, Meguro M, Yamadori A, Sekita Y. Neuropsycho-epidemiologic study on aging and dementia in a community: The Tajiri Project. *Research and practice in Alzheimer's disease*, Serdi Publisher, Springer Publishing Company, NY, 2003, pp 80-85.

Meguro K, Community-based Measures to mild cognitive impairment (MCI): The Osaki-Tajiri Project. *Psychogeriatrics* 2007;7:132-136.

Meguro K. *A clinical approach to dementia: An instruction of CDR worksheet*. Tokyo: Igaku-shoin, 2004 [in Japanese].

Meguro K, Ishii H, Yamaguchi S, Ishizaki J, Sato M, Hashimoto R, Meguro M, Lee E, Tanaka Y, Kasuya M, Sekita Y. Prevalence and cognitive performances of Clinical Dementia Rating 0.5 and mild cognitive impairment in Japan: The Tajiri Project. *Alzheimer Disease Associate Disorders* 2004;18:3-10.

Meguro K, Ishii H, Kasuya M, Akanuma K, Meguro M, Kasai M, Lee E, Hashimoto R, Yamaguchi S, Asada T. Incidence of dementia and associated risk factors in Japan: The Osaki-Tajiri Project. *Journal of the Neurological Sciences* 2007;260:175-182.

Meguro K. International report: Local response following the Great East Japan Earthquake 2011. *Neurology* 2011;77:e12.

Meguro K, Tanaka N, Kasai M, Nakamura K, Ishikawa H, Nakatsuka M, Satoh M, Ouchi Y. Prevalence of dementia and dementing diseases in the old-old population in Japan: The Kurihara Project. Implications for Long-Term Care Insurance Data. *Psychogeriatrics* 2012a;12:226-234.

Meguro K, Tanaka N, Nakatsuka M, Nakamura K, Satoh M. Vascular lesions in mixed dementia, vascular dementia, and Alzheimer's disease with cerebrovascular disease: The Kurihara Project. *Journal of the Neurological Sciences* 2012b Aug 4 [Epub].

Meguro K, Kasai M, Akanuma K, Meguro M, Ishii H, Yamaguchi S. Donepezil and life expectancy in Alzheimer's disease: A retrospective analysis in the Tajiri Project. *BMC Neurology* 2014a;14:83 [Epub].

Meguro K, Suzuki R, Nakata E, Ishii H, Yamaguchi S. Cognitive impairment affects mortality in older people in Japan. The Tajiri project 1991-2005. *Journal of the American Geriatrics Society* 2014b;62:1793-1795.

Morris JC. The Clinical Dementia Rating (CDR); Current version and scoring rules. *Neurology* 1993;43:2412-2414.

Mussweiler T. Comparison processes in social judgment: mechanisms and consequences. *Psychological Review* 2003;110(3):472-89.

Nakamura K, Kasai M, Ouchi Y, Nakatsuka M, Tanaka N, Kato Y, Nakai M, Meguro K. Apathy is more severe in vascular than amnesitc mild cognitive impairment in a community: The Kurihara Project. *Psychiatry and Clinical Neurosciences* 2013;67(7):517-525.

Nakano M, Kondo M, Wakayama Y, Kawana A, Hasebe Y, Shafee MA, Fukuda K, Shimokawa H. Increased incidence of tachyarrhythmias and heart failure hospitalization in patients with implanted cardiac devices after the great East Japan earthquake disaster. *Circulation Journal* 2012;76:1283-1285.

Neria Y, Sullivan GM. Understanding the mental health effects of indirect exposure to mass trauma through the media. *Journal of th eAmerican Medical Association* 2011;306:1374-1375.

Normile D. Japan disaster. Scientific consensus on great quake came too late. *Science* 2011;332:22-23.

Ogawa S, Ishiki M, Nako K, Okamura M, Senda M, Sakamoto T, Ito S. Effects of the Great East Japan Earthquake and huge tsunami on glycemic control and blood pressure in patients with diabetes mellitus. *British Medical Journal Open* 2012;2:e000830.

Rabinak CA, Nirenberg MJ. Dopamine agonist withdrawal syndrome in Parkinson disease. *Archives of Neurology* 2010;67:58-63.

Román GC, Tatemichi TK, Erkinjuntti T, Cummings JL, Masdeu JC, Garcia JH, Amaducci L, Orgogozo JM, Brun A, Hofman A. Vascular dementia: diagnostic criteria for research studies. Report of the NINDS–AIREN International Workshop. *Neurology* 1993;43:250-260.

Satoh M, Kikuya M, Ohkubo T, Imai Y. Acute and subacute effects of the great East Japan earthquake on home blood pressure values. *Hypertension* 2011;58:e193-194.

Satomi S. The Great East Japan Earthquake: Tohoku University Hospital's efforts and lessons learned. *Surgery Today* 2011;41:1171-1181.

Schultz RR, de Castro CC, Bertolucci PH. Memory with emotional content, brain amygdala and Alzheimer's disease. *Acta Neurologica Scandinavica* 2009;120:101-110.

Silove D, Steel Z, Psychol M. Understanding community psychosocial needs after disasters: Implications for mental health services. *Journal of Postgraduate Medicine* 2006;52:121-125.

Simon H. Altruism and Economics. *AEA papers and proceedings,* May 1993, Vol 83, No 2.

Takada J, Meguro K, Sato Y, Chiba Y; Kurihara Project Members. Life concerns of elderly people living at home determined by Community General Support Center staff: Implications for organizing more effective integrated community care system. *The Kurihara Project. Psychogeriatrics* 2014;14(3):188-195.

Takeda M, Tanaka N, Chiba T. Prospects of future measures for persons with dementia in Japan. *Psychogeriatrics* 2010;10:95-101.

The Alfred E. Alquist seismic safety commission to the California state legislature. *The study of household preparedness: preparing California for earthquakes.* 2009.

Yamaguchi S, Meguro K, Ishii H, Meguro M, Akanuma K. Assessment of mental deterioration with the Cognitive Abilities Screening Instrument (CASI) and glucose hypometabolism in Alzheimer's disease: The Osaki-Tajiri Project. *Journal of Clinical Neuroscience* 2009;16:1430-1434.

Yamashita. *Tsunami-tendenko: A history of tsunami in Modern Japan.* (Tsunami-tendenko: kindai nihon no tsunami shi). [Japanese] Shin-Nihon Pub Co., Tokyo, 2008. ISBN 978-4-406-05114-9 C0044.

Yeung NC, Lau JT, Yu NX, Zhang J, Xu Z, Choi KC, Zhang Q, Mak WW, Lui WW. Media Exposure Related to the 2008 Sichuan Earthquake Predicted Probable PTSD Among Chinese Adolescents in Kunming, China: A Longitudinal Study. *Psychological Trauma* 2016 Jul 18. [Epub].

ACKNOWLEDGMENT

I express my deepest sympathy to those suffering from the earthquake and offer my sincerest condolences to the relatives of the victims.

AUTHOR'S CONTACT INFORMATION

Professor Kenichi Meguro, MD, PhD
Geriatric Behavioral Neurology,
Tohoku University CYRIC
Email: k-meg@umin.ac.jp

INDEX

E

F

G

H

I

J

K

M

N

O

P

S

T

V

W